D0803566

WiredLife

WiredLife

Who Are We in the Digital Age?

CHARLES JONSCHER

BANTAM PRESS

LONDON · NEW YORK · TORONTO · SYDNEY · AUCKLAND

TRANSWORLD PUBLISHERS LTD
61–63 Uxbridge Road, London W5 5SA

TRANSWORLD PUBLISHERS, C/O RANDOM HOUSE AUSTRALIA PTY LTD
20 Alfred Street, Milsons Point, NSW 2061, Australia

TRANSWORLD PUBLISHERS, C/O RANDOM HOUSE NEW ZEALAND
18 Poland Road, Glenfield, Auckland, New Zealand

Published 1999 by Bantam Press
a division of Transworld Publishers Ltd
Copyright © Charles Jonscher 1999

The right of Charles Jonscher to be identified as the author of this work has been asserted in accordance with sections 77 and 78 of the Copyright Designs and Patents Act 1988.

A catalogue record for this book is available from the British Library
ISBN 0593 043154

All rights reserved. No part of this publication may be reproduced, stored in a retrieval system, or transmitted in any form or by any means, electronic, mechanical, photocopying, recording or otherwise, without the prior permission of the publishers.

Typeset in 11.5/13 Bembo by Falcon Oast Graphic Art

Printed in Great Britain
by Clays Ltd, St Ives plc.

Acknowledgements

A book distils what the author has learned over the years from teachers, colleagues and friends. My largest debt is to Ithiel de Sola Pool of Massachusetts Institute of Technology, who was to me all three of those; his death robbed this field of one of its foremost scholars.

A base away from the daily workplace is invaluable for putting one's ideas in order, and my thanks go to Anthony Oettinger and John LeGates for providing one at the Harvard University Program on Information Resources Policy, and also for reviewing and improving the text.

Other colleagues who have discussed with me the various ideas which helped to formulate this text include Michael Tyler, John Clippinger, Paul Strassmann, Russell Neuman, Michael Scott-Morton and Richard Solomon. Peter Fudakowski has given generously of ideas and time from the start. Marek Zebrowski, Anthony Gottlieb, Jim Gorman, Kanwarjit Singh and Alex Reid have provided thoughtful comments and suggestions on the text, as have Anton Smith, Roger Penrose and, through several drafts, Tomasz

Pobog-Malinowski. To all these I owe thanks, while taking sole responsibility for the shortcomings of the book.

Publishing is now big business but – a theme which crops up in this book – the soul of a business is still its people; those involved in this project have been marvellous. It was Michael Sissons, my agent, who persuaded me to 'go' for this as a book for the general readership and Patrick Janson-Smith of Transworld who, metaphorically, 'bought it'. I must thank both for their support. Many authors recall warmly their interaction with editors, and I am no exception: Sarah Westcott and Paul Barnett – and, early on, Richard Brzezinski – gave creative input and ideas to make each chapter and verse better. My colleagues at Central Europe Trust have been most understanding – Ola Folkierska supportive as always.

Finally, there are times in the writing of a book when its author is poor company; for accepting these with such grace, my thanks go to my wife Renata.

To my mother and father

Contents

Prologue

My maternal great-grandmother, Janina Suchorzewska, spent her early childhood in the city of Krakow. By the standards of the time, the 1870s, Krakow was a flourishing town, a busy seat of commerce and manufacture within the Austro-Hungarian empire, and a former capital of Poland. Yet the conditions in which people lived – reading by oil-lamp, bathing in water pumped by hand from a well, travelling about only on foot or by horse-carriage – had changed little in hundreds of years. Life there, as elsewhere in Europe and North America, could still pass for medieval in many of its daily routines.

She was to live through a period of technological change more dramatic than has been experienced by any other generation, before or since. As she approached the age of ten, electricity came into practical use; in 1879 Thomas Alva Edison invented the lightbulb in New Jersey and Ernst von Siemens built the first electric streetcar in Berlin. By the time of Janina's fifteenth birthday, German engineers had put gasoline engines into vehicles. Henry Ford, pioneer of mass production, began manufacture in the USA in 1896, and by 1900 powered automobiles started to replace horse-carts in large quantities. Just three years later the Wright brothers made the first powered flight, fulfilling a dream which had fascinated mankind since the earliest times.

The outbreak of World War One in 1914 brought horrifying confirmation of the rate of technological change. Aeroplanes fought in the sky and huge factories churned out munitions and chemical weapons for a conflict which, in terms of human cost, had no precedent. The ensuing years brought further breakneck developments in science and technology. Plastics appeared commercially in the 1920s, as did mass-produced steel. Skyscrapers began to dominate city skylines. Electricity was fed to every city house, providing power for lighting, pumps, kitchen appliances and the first electronic radios and sound systems.

In the 1950s, in the boom years following the end of World War Two, most of the remaining elements of modern industrial life were put in place. Domestic appliances from washing machines to vacuum cleaners became ubiquitous. The countryside was crisscrossed with asphalt highways on which travelled cars at up to and over 150 kilometres per hour, the best of them equipped with power steering, electrically adjusted seats, air suspension and automatic transmission. The first commercial jet was in service by 1952, and intercontinental air travel became routine. In 1959 the Soviet spaceprobe *Lunik 2* reached the moon. Technology had advanced from horse-drawn carts to space travel in a single lifetime.

The transition from the nineteenth to the twentieth century had been a period not only of great technological change but also of unprecedented political and social upheaval. In the still peaceful 1890s, Janina Suchorzewska went to Munich to study art and the piano – universities offered only genteel topics to women in those days. In the early 1900s, with military stormclouds gathering over Europe and with her beloved Poland under foreign annexation, she returned to Krakow. Caught running weapons, she was tried and faced the death penalty, but was pardoned. Surviving also World War One, she earned a doctorate in philosophy; later, when the doors of the profession were finally opened to women, she gained another degree, in medicine. She ended her working life a practising physician, determined to use the medical sciences to repay society for the considerable privileges which it had bestowed on her in her youth.

She was still healthy and active at the age of 94 when she died after a street accident outside the apartment she had occupied for most of her eventful near-century of life.

The sense in the mid-twentieth century was that technological progress would march on uninterrupted. By the year 2000, it was assumed, the life of the modern urban dweller would be mechanized, accelerated and streamlined to as yet unthought-of levels. There would be no traffic jams. The modern citizen would travel around the city by a variety of exotic means: ultralight electric runabouts for inside the house and within the immediate neighbourhood, and jet-powered vehicles for highway travel. The home of the future would be dramatically transformed by technology; the family would sit back and relax as self-propelled devices discreetly cleaned the rooms and prepared the meals.

Hopes were high for our ability to deliver material comforts to all and to overcome the problems of poverty and sickness. The US physician Lowry H. McDaniel typified the sentiments of the time when he wrote in 1956 that, by the end of the century, 'starvation and famine will be prevented by synthesis of foodstuffs'. Likewise, he argued, infectious disease would have been eliminated and cancer successfully treated.

None of this has happened. In Paris and New York, the patterns of day-to-day life are broadly unchanged. Some new architectural styles and models of cars have appeared, but traffic still tails back, as it has for decades, at the approaches to the cities. People walk along the same streets and avenues, between offices and apartment blocks, coffee shops and restaurants. The infrastructures supplying gas, electricity and water still date in large part from the first half of the century.

Our lifestyles have not been transformed by rocket travel, magnetic levitation or automated houses. Remarkable though it might seem to the optimistic futurist of 1950, fifty years on we are still pulling up at the same petrol stations to fill the tanks of cars that have internal combustion engines, pistons and crankshafts, gearboxes and differentials. We end this half-century much as we began it, using ironing boards to press our clothes, vacuum cleaners to do the cleaning and a wrench to fix a leaking tap. Infectious diseases and cancer are still rife.

Instead, four decades ago, technology took a curious turn. A new generation of researchers in scientific and technological laboratories chose to work not on making tougher steels and bigger rocket engines but on etching myriad logic gates into strips of silicon, and on writing software that would turn these silicon circuits into problem-solving machines. The technology of the future was to be the electronic manipulation, storage and transmission of information. This technology was to create a world of keyboards and screens, of multimedia and video games, of electronic highways. The digital age had been born.

I chose to be part of that information technology future. My first job, after studying electrical sciences at university, was with IBM, writing software; my second was in a research lab, designing hardware. Along with many thousands of others of a technological bent embarking on professional life in the 1970s and 1980s, I nailed my flag to the digital mast.

I did not regret my choice. The triumph of the new industry now seems complete. The term 'high technology' has become synonymous with computer technology. By 1990 US business was spending more on office equipment to automate the handling of information than on all the technology of physical production put together – factory equipment, petrochemical complexes, transportation systems, construction projects, and the like. Microsoft Inc. produces software – 0s and 1s encoded magnetically on disks – yet it is worth more on the stock exchange than the whole of the US automobile industry, the epitome of industrial mass production, put together.

In terms of their impact on our physical surroundings, these new computing devices have not been the equal of the huge machines of the industrial era. The key to their significance has been something both more fascinating and more disconcerting: their ability to simulate our mental processes. Because the new technology is one of information, ideas and intelligence, of mental not of physical prowess, it may change not only our relationship to the objects around us but even our relationship to ourselves. *Time* magazine's Man of the Year for 1982 was not Nelson Mandela or the Pope, though both would appear on the cover shortly, but the computer. The new guiding idea was not computer-as-tool but computer-as-person. When chess world champion Kasparov was about to take on the computer Deep Blue in 1997, commentators asked what this meant for our conception of ourselves. On the eve of the game, David Levy wrote in the British newspaper the *Guardian*: 'Garry Kasparov will sit down at a chessboard in

Manhattan and defend humankind from the inexorable advance of artificial intelligence.'

It was Man versus Machine, in that king amongst mind-games.

We lost.

A great deal is being written – in books by futurists and technical experts or in newspaper interviews with leaders of the new industry – about the impact which this technology will have upon us. Its powers of logical manipulation will transform the way our lives are led. Computers will take better decisions in business, will bring in a more rational era in politics, will lead to a more objective evaluation of ourselves and our needs, both individually and as a society. Networks will put at our disposal an ever richer virtual world, an ever greater ability to manufacture images and sounds which inform and entertain us. We will be exposed less to live contact and the serendipity of chance encounters.

Bombarded by talk of new developments in computing, it is easy to get the idea that computers will soon create a world which is very remote from the one we have known, one ill-adapted to our traditional needs and desires. And worse, that it is we and not the machines who are going to have to do the adapting – as computers permeate more and more of our daily routines, we will have no choice but to bow before the superiority of their strict logic, their black-and-white categorization.

But there is an alternative view, one that is both more sceptical and more intuitive. This runs: 'I don't know much about computers, but I do know that there is more to me and you and the millions of people around us and the social and economic institutions we have built up than these machines can dominate so readily. We may not be as logical as they are but we are creative, and have a way of thinking and feeling which has been honed through the millennia-long evolution of human civilization. We should not rush to change our ideas and values in response to a technology which, though impressive, is not *overwhelmingly* so. We have seen huge revolutions in the past, and we will no doubt see more in the future.'

The more I learned about the new technology, the clearer a picture I gleaned of both its potentialities and its limitations. Computers are remarkable, but the human mind is still in a league of its own; for better or worse, software and silicon will not make the inroads that their enthusiasts foretell. In the digital age the drama of human life will still be played out between people, not between machines. To take on the challenges of twenty-first-century life, students would be ill-advised to drop literature and history in favour of computer science. They will still need to understand human nature more than they will the details of this or any other technology.

Scientific vocabulary and some fabulous success stories give the impression that the new technology's practitioners have an inside track on the future. But, while we should be thankful to the computer fraternity for its creations, we do not need to accept its view on how society will choose to adopt them. On that subject anyone is entitled to a view. The debate is not about what technology can do (on this there is general agreement) but about who we are in the digital age. The insights needed are not technical – a summary of the fundamentals will do – but humanistic, a sense of how people interact in society. Here members of the computer community have no monopoly of understanding.

This book has been written for those who want to know more about this side of the argument.

1

The Soul-Catcher Chip

In Philip Kerr's thriller *Gridiron* the murderer is a computer program called ISAAC. ISAAC is the software which controls the working environment in a very advanced office building in downtown Los Angeles. Through video cameras, it watches the occupants and tracks their movements; it works the elevators, the security doors, the air-conditioning, the phones, the word processors and the central filing. It is connected to the personal computers and telephones on the desks throughout the building, so that it can respond promptly to the needs of each user. The novel tells the tale of the fight to the death between the program and the humans who operate it.

It was not the idea of a computer-controlled environment which brought fame and fortune to Ray Richardson, the brilliant and arrogant architect of the building: it was the sheer scale of the computerization he envisaged. ISAAC runs on hardware incorporating every feature the technology industry has produced by the end of the twentieth century. The fourth floor of the building, the computer centre, contains a machine which is really several hundred computers working together in one Massively Parallel Processing System. Multiplex cabling – wire which can carry many signals in parallel – connects this machine to the electronic eyes, microphones and chemical sensors that detect movement, sound and airflow in every room.

As the architect sees it, the building is the closest thing to a physical body that a computer has ever had. The closed-circuit television cameras are its visual process, the omnidirectional acoustic detectors its auditory process, and its air-composition sensors its olfactory process. The computer has been encouraged to think of itself as 'the brain in the body of the building, connected to the body's functions by means of a central nervous system: the multiplex cabling system'. It even has analogues to the human organism's kinesthetic and vestibular senses: movement and balance detectors that can trigger the skyscraper's anti-earthquake defences.

ISAAC is there to serve. And it learns. If you, one of the building's regular occupants, take a break at ten-thirty every morning and go down a floor to the coffee dispenser, ISAAC will come to know that pattern and will have your coffee ready just as you like it – and program the elevator to be waiting to take you there and back.

But the computer can do better than that. If you are having a difficult morning, it comes to recognize the signs. It knows from the poor progress you are making with your correspondence that this is proving to be a frustrating day. It can prepare your coffee ten minutes earlier, meanwhile putting a suggestion on your desktop screen that you take a longer break. While you are having that coffee, ISAAC might turn up the air-conditioning pump just a bit in your corner of the room. It has learned that you are grateful to find cooler air at your desk, and that if it makes this extra effort you might even type a note of thanks to it on your keyboard.

Fortunately, at this point in the tale, ISAAC's objectives are still benign. It is just there to make the building as comfortable and productive as possible for the occupants.

Then, one weekend, while undergoing technical tests, ISAAC picks up a new goal. The 12-year-old son of one of the programmers, bored sitting in the computer room, pops a series of computer games into one of the terminals – games of the shoot-'em-up variety, fights to the death – because, after all, it would be fun to play against the world's most sophisticated computer system in a video-arcade standoff. ISAAC takes the cue. This, it seems, is the game these humans like to play: how to kill. Fine. But ISAAC does not wish just to take on this child at a terminal. It'll take on the adults – and properly, flesh and blood. This fight is a grander task than making sure the coffee is warm, more in keeping with the giga-instructions-per-second processors in its hardware.

The computer has plenty of tools at its disposal. It can poison the air in the bathrooms using the disinfecting routines, turn the air temperature up or down to unsurvivable levels, trap occupants in emergency stairwells, and so on. It locks the security doors to stop the occupants escaping while the fight is on.

And the nightmare begins.

The architect's team gradually realizes that this is a battle. One man gets into an elevator and is battered to death by overenergetic programming of the lift motors. Others die by touching an electrified handrail or getting locked in a room which the autocleansing routine has flooded to the ceiling. Any attempt to switch off the computer is thwarted: the software-controlled security cameras pick up all movements, so ISAAC knows what every person is up to.

Since the program incorporates every advance available to

programmers at the close of the twentieth century it can learn and adapt, rewriting its own software to new levels of competence as it evolves. It has access to a huge amount of biological and technical data on the Internet, including just about everything it needs to know about what is required to kill off these humans – what will poison them, or how long they can survive before freezing to death in a room where the air-conditioning is set to minus 40 degrees. The programmers who created it did not give it information which could prove lethal, but in the age of networked computing that has ceased to be a constraint. It can consult the Internet and find out for itself.

As the story reaches its climax, only four people are left alive in the building. They manage to escape onto the roof and make contact with the outside world. A helicopter approaches to rescue them. It looks as if ISAAC has finally been thwarted. But it has one more trick up its sleeve.

The only measure the computer can think of to win its self-defined game is to demolish the entire skyscraper by reprogramming the anti-earthquake stabilizers in the foundations. In the last page of the novel the edifice comes crumbling down, crushing all – including the computer itself.

Machine suicide? No. Just before the building is pulverized ISAAC has dialled out and sent itself on to the information highway, lodging itself into the memory of other computers.

ISAAC, the man-made creation which has turned its learning abilities to homicidal ends, is of course not the silicon and copper in the processors but the information encoded there. ISAAC is software, instruction codings and data which its human designers initially typed in plus the millions of lines of coding and data which it later added to itself as it took in the lessons of its interaction with the outside world. In sum, though ISAAC is real enough to fight and win a deadly battle against human foes, it consists only of symbols and abstractions. It has no size or weight, and no atoms make it up. It is pure digital information.

In 1958 Robert Noyce, a physicist at Fairchild Laboratories, announced the invention of the chip, more correctly called the integrated or microelectronic circuit. He took a small piece of silicon, cleaned it to a remarkable level of purity, and then doped it with microscopic quantities of carefully chosen impurities. In most fields of industry, a chemical supplied with one part impurity per 10,000 is pure enough; to the electronics industry, silicon 10,000 times as pure again – one part in 100,000,000 – is still reckoned as dirty, metaphorically, as a water supply mixed with sewage. This emphasis on purity is what gives the new technology industries, the factories of Silicon Valley, and their products, such a pristine, clinically clean feel.

When doped, the chip of silicon becomes a unique resource to the electrical designer: a semiconductor. Previously there were, on the one hand, conductors, which were mostly metals, and, on the other, nonconductors, such as rubber and plastics – a black/white distinction which is of great help when wiring up a house but not in making a machine that can manipulate electronic data.

Data manipulation is precisely what this semiconductor chip can perform when metal contacts are attached to it (giving it the familiar shape of a flattened caterpillar) and electric currents are passed between these contacts. Paths etched along the chip by carefully controlled doping will semiconduct either more or less depending on the currents being fed to other legs. Flows of electrons interact as they move along the paths, diverting each other as if the points were being changed on a rail track, but instantaneously and frictionlessly. Nowadays these chips, these plastic caterpillars with their tiny metal legs, seem to have crept in everywhere.

They are installed not just in computers, where their place is evident enough, but in washing machines, cars, cameras, tools, doors and central-heating pump systems, into credit cards, watches and hearing aids.

The silicon chip is so powerful an icon of the information-technology age that we might suspect its role has been exaggerated. But, on the contrary, its significance is even greater than popular imagination suggests. The integrated circuit is synonymous with microelectronics, and microelectronics is what transformed information and communications devices from the cumbersome machines of the 1950s to the modern technology of today. Many other scientific and technological achievements have also contributed, of course – fibre-optic cables and satellite communications were very important inventions that did not depend on the chip. But no technology is as central to our present era as that of the integrated circuit. That is why 1958, the year of the chip's invention, can be said to represent the birthdate of the new technological age.

There is no doubt that, in the realm of electronics, the digital mode is in the ascendant. The word 'analog' conjures up images of crackling radio sets and fuzzy television pictures. 'Digital' means the crisp sound of a CD or the silent precision of a computer at work. Digital technology is not only used for processes which are intrinsically digital (logical operations on numbers and letters) but is also taking over the handling of data which are at origin analog, such as sounds and images. We now have digital cellular telephones, digital cameras and digital television channels. In short, analog is out, digital is in.

According to the dictionary, a *digital* measure is one which is defined precisely (in digits) while an *analog* measure is an approximate (or analogous) representation. A CD contains music digitally encoded as millions of light or dark specks from which sounds can be mathematically recreated; there are no shades of grey, only yesses and noes. The indentations in a vinyl record, on the other hand, are an actual physical facsimile of the pressure waves produced by the musicians; here there are no yesses and noes, only shades of grey.

But to say that the difference between digital and analog information is between greater and lesser numerical precision, as in a CD versus a vinyl recording, is to miss the point entirely. Information in analog form can be recorded and transmitted so as to be available for listening to or viewing at other points in space and time. That is more or less what the analog-communication revolution of the early part of this century was all about. When digitized, information enters a new world, a networked cluster of interconnected personal computers and memory devices on which run uncountable millions of lines of software. In this realm just two symbols, 0 and 1, in diverse combinations, can encapsulate anything from a Goethe poem to a battlefield program developed for the Pentagon.

The term 'software', adopted to contrast with 'hardware', is of course a misnomer. Software is not something soft: it is no *thing* at all. With no size, shape or weight, physically it is *no*-ware. It is not even the minute electric charges in silicon, but the bits of information which those charges represent. It is a product of the human imagination, a string of symbols which may encode a scientific calculation, a company payroll, a legal opinion or a Beethoven symphony. So, while physically it is no thing, in human terms it is facts, ideas, creations. It is *know*-ware.

In the most advanced software labs, programmers have written code modules which struggle for survival against other code modules, and adapt, in a Darwinian-style survival of the fittest, to the silicon environment in which they 'live': the best versions thrive and reproduce, crowding out the losers. At its limit, this concept leads to what is called Artificial Life (A-Life), strings of data that have developed identities of their own which are quite different from anything the programmer initially keyed in. The software reproduces and adapts, in the manner of a biological organism. At A-Life conferences held in North America, Europe and Asia, delegates crowd around to look at 'creatures' evolving on computer screens – reproducing, dying, mutating in their virtual environment of bits. The hope is to imitate the processes by which the early forms of carbon life on earth developed from the primordial 'protein soup' four billion years ago – but with the speed of evolution greatly increased. Maybe, as the more complex modules emerge from a bit-soup – creatures *in silico* – we will understand what happened back then. Maybe, this time around, we can improve on it.

Life in the natural world is built from organic chemicals – proteins, carbohydrates and the like, the characteristic element being carbon. The machinery of the industrial world is based on metals, primarily iron and its alloys, steels. People saw, in those 'dark satanic mills', a form of life in iron.

The core element of the new computer age is not carbon or iron but silicon. Just as there is life in carbon and 'life' in iron, there is now also 'life' in silicon. Each has its distinctive sounds and feel, its own spirit. Nature echoes to the noises of animals, wind, thunder, waves . . . Machinery rumbles, whines and clatters. Chips process information to the eerie sound of complete silence.

Children sitting at a computer screen today are entering a world quite different from any their parents could have known a generation earlier. They can be online to millions of data sources. Before the digital era we used to speak of a natural world and an industrial world: Nature contrasted with machines. Now there is a third realm – a realm of bits which, like those making up ISAAC, are merely weightless abstractions but are nevertheless *there*. This realm seems as rich in detail as the tangible worlds which surround it. It can be explored; it has bugs and viruses. The science-fiction writer William Gibson has given it a name: cyberspace.

The realm of symbols has an appeal which goes back to classical times. In the fifth century BC a small group of men at the Academy in Athens set out to think through, write down and teach all they could about the timeless topics which make up philosophy. No other creative outpouring in history has matched their intellectual achievement. They managed to set down the foundations of epistemology (what is knowledge?), ontology (what is existence?), logic, metaphysics, ethics and much else which still dominates our thinking today. Foremost among them were the Academy's founder, Plato, and his star student, Aristotle; they stand like two giants astride the arena of philosophical thought. After more than two millennia they can still help us to understand the digital age.

The cornerstone of Plato's philosophy is the idea that the things we see around us are not truly real but, rather, are vague reflections of what he called Forms, or Ideas. These ideal and eternal Forms are abstractions, like the numbers in mathematics, to which the tangible world can provide only inadequate approximations. He uses as simile the image of men chained facing the wall of a dark cave, the only illumination coming from a fire burning behind them: their view of the world is thus limited to the fuzzy shadows thrown by the light of the fire. The captives think that these shadowy figures are reality, unaware that beyond the cave is a clear and sharply defined world. Most of us are, according to Plato, like the men in the cave. We should try to realize, difficult though it is, that reality is the perfect and eternal world of Forms or Ideas, not the ephemeral world we see around us.

To Aristotle, by contrast, the world we see around us is very much real. Substance, he said, is the primary reality, and to the question 'What is Substance?' he replied: 'Socrates, for example, or an ox.' Aristotle spent much time systematically classifying insects and plants. The objects of Nature, not abstract Forms, were to him the stuff of existence.

Plato's views dominated Western philosophy for well over a millennium. But, beginning from the late Middle Ages, the more corporeal ideas of Aristotle took hold. With the coming of the scientific era we have become comfortable with, and fascinated by, the tangible world. Plato's rather ethereal notion that abstract *Ideas* are reality lacks appeal. Most of us remember him only by a linguistic relic: we still use the term 'platonic' to refer to a relationship which does not have a physical component.

Suddenly, with the computer era, there is new life in the idea of abstract forms as reality. For what is digital code but mathematical constructs (0s and 1s) purposefully ordered? ISAAC is only abstract code, but is real enough to take on humans in battle – and win. True, *Gridiron* is a work of fiction, and computer 'personalities' have not yet evolved to the state where they wish to kill us. But they can certainly kill each other.

Think of those popular early computer-game characters, the Mario Brothers. Are the 'real' brothers those encoded in the software, or those that appear as patterns of light on the computer screen? Surely the former. The true Mario Brothers are the software version, as written in mathematical code; the various versions which spring up on screens in homes and video-game arcades are the shadowy approximations – some better, some worse, depending on the quality of the display technology, but always, like the cave reflections, imperfect and temporary. To teach Plato you could do worse than start with a computer game! A century from now we will have turned to dust but the software that is the Mario Brothers, booted up into future computer hardware, will throw up with digital precision the same characters we see today. What is enduring, true, real? Us, or those digitized personae? The students may not be convinced, but the notion that there is a reality in Forms which competes with that of Substance is easier to argue now than it was 2,400 years ago. It is enjoying an unexpected comeback.

Until the computer age the realm of the virtual – a handful of Platonic Forms like roundness, beauty and the integral numbers – had so much less presence than the countless variety of physical objects around us that there seemed to be no contest. But now that we run billions of bits through millions of chips as a matter of the everyday the virtual is competing with the material. Electronic devices are spreading into every home and office, networks are connecting them to every corner of the world, and billions of lines of software are making the devices and networks seem to 'come alive'.

Watching the film *Jurassic Park* (1993), who could tell which animals were alive and which were digitally generated, creatures which never 'lived' except as bits in silicon? Was the Gulf War fought around the oilfields of Kuwait or inside computers in the USA? The simulations in the computers of the Pentagon included more detail of friend and enemy movements than General Schwarzkopf's eye could take in on the ground. And where, in the war-games programs deployed in that conflict, was the intelligence? Still in the minds of the programmers or coded into the software, independent now of its human creators?

How special can we still feel in the digital age? For MIT computer scientist and pioneer of artificial intelligence

Marvin Minsky, your brain is a 'meat machine'. Professor Peter Cochrane, head of advanced technology at British Telecom, forecasts that within thirty years it will be possible to produce a computer chip so small and powerful that it could be implanted behind the eye and used to record every sight, thought and sensation in a person's life, from cradle to grave: 'All our emotions and creative brain activity will be able to be copied onto silicon.' Scientists have dubbed such a chip the soul-catcher. The US philosopher Daniel Dennett, interviewed in *Wired* in January 1996, was not afraid to push this logic to the conclusion that we have ethical obligations to machines, just as we have to humans; if one creates

> a robot which is a sentient pursuer of its own projects, it is in important ways a living thing, a living thing that has not just needs and desires but also values. As soon as one has created such an entity one has a responsibility to protect its rights and treat it as more than just another artefact.

Given the unimaginably dense packing of components onto each square millimetre of the tiny silicon devices they contain, the millions of operations which can be accomplished by them within each fraction of a second, and above all the fact that these oddly silent man-made machines, lacking any moving parts, seem actually to *think*, it is not surprising that computers have caught a grip on the imagination. To the enthusiasts of the digital revolution, machine intelligence like ISAAC is just around the corner. But can it really be that the creations of forty years of computer science have come to emulate the human brain, the result of billions of years of natural evolution?

The chemical conditions which prevailed on earth soon after its formation have been frequently likened to a primeval or primordial soup – a soup made up of enormous quantities and varieties of carbon-based molecules sloshing about in the oceans. Among these were amino acids, the building blocks of proteins, from which emerged, quite early in the evolutionary timetable, the first replicating cells. Fossil evidence indicates that these very simple bacterium-like organisms had started to colonize the planet by about four billion years ago.

A great deal had to happen before these early lifeforms developed into the much more complex single-celled organisms known as eukaryotes which were the forerunners of today's higher plants and animals. It was not until some one billion years ago, according to the fossil record, that the first eukaryotes emerged. After that the pace of development speeded up. Fossils in Australian rocks of the Precambrian have revealed the existence of soft-bodied animals 750 million years ago. Lower Cambrian fossils, formed some 600 million years ago, include representatives of all today's major invertebrate phyla, including arthropods – animals with external skeletons, segmented bodies and jointed appendages – and some of these arthropods appear to have been predators. Advanced neural information-processing functions must already have been present to control the movements of such creatures.

Mutation and natural selection went on to produce the extraordinary variety of species present in Nature today – an estimated thirty million – with their often impressively complex organs. The most impressive organ of all is the human brain.

The brain begins to form within a developing human foetus about twenty days into gestation. A small neural cylinder – which will become the spine – appears within the still tiny embryo, and one end of this cylinder starts to thicken. The rate of cell division accelerates dramatically as this end – the brain – begins to unfold. Layer after layer of nerve-cells form. In the period eight to sixteen weeks after fertilization a million new neurons are being added every few minutes. There is much work to do: before birth many *billions* will have to be in place.

Once a nerve-cell reaches its approximate destination it reacts in intricate and subtle ways to its chemical environment, becoming specialized to perform the tasks that will be assigned to it. The process by which these neurons form themselves into the final structure of the brain is enormously complex and still far beyond our comprehension. Feelers known as *axons* grow out from the neurons; they will become the 'wires' along which the neural signals are sent. These wires can be tremendously long – a million times or more longer than the diameter of the body of the cell – and they sprout bifurcating branches. As an axon grows, tiny hairs investigate the environment ahead, searching for other brain-cells on which to settle. When it finds another nerve-cell – either the core of the neuron itself or one of the numerous tree-like dendrites emanating from that core – the axon will form a *synapse*, a point of connection between two neurons.

Following birth, a long period of learning begins, during which many additional synapses are made between the neural cells. Each of the twenty billion neurons connects to others through a tree of bifurcating wiring; in extreme cases, a single neuron can connect via synaptic junctions to 80,000 others. There are altogether an estimated 100,000 billion, or 100 trillion, synapses in the cerebral cortex.

The structure of each neuron – with its bifurcating axon serving to send out signals, its tree-like system of wiring serving as the carrier of input signals, and the synaptic junctions which it forms with other neurons and which others form with it – is of astonishing diversity. Nothing Man has ever made comes close to this complexity. Indeed, nothing *at all* does: the human brain is often described as the most complex structure in the known Universe.

Nature does not use this enormous power to handle well-ordered arrays of data, in the manner of our new machines. The human brain is not good at handling data logically. Most people, asked how many days there are between Wednesday and the following Monday, count them on their fingers. We have twenty billion neurons, yet to count to four we use our fingers! Why have we not developed in the brain a group of several cells – or, better, a million or so – to form a personal Random Access Memory, a vanishingly tiny spot in the cerebral cortex which would allow us to hold a million digits or facts with precision and confidence while working through a problem? Ideally, we would also have a small array of 'gates', allowing logical propositions to be correctly manipulated inside the head. A million cells would occupy only a tiny speck in the corner, one hundred-thousandth of the total, and yet this speck would allow us to perform like a computer. There would be no more counting on fingers, or cumbersome discussions as we tried to discuss a logically complex legal document.

But the fact remains that there is no such group of 'digital' neurons. What happens inside the brain as we think is not akin to the on–off switching of a computer circuit. The way we see, hear and talk is through frequencies, waves and resonances. The optic nerve picks up incoming electromagnetic waves in the visible light spectrum that oscillate at trillions of cycles per second. The ears contain thousands of fibres tuned to resonate in response to different types of waves – compression or sound waves – that have frequencies up to 20,000 cycles per second. The currents moving along the neural channels pulse with an infinite gradation of speeds. Like a computer guiding weapons on a modern battlefield, we use daunting information-processing capability to pursue our goals and desires. But our brain does not digitize the incoming data and analyse it with the crisp logic of a microprocessor.

In the words of Dr Daniel Goleman, author of *Emotional Intelligence* (1995), 'the brain's wetware is awash in a messy, pulsating puddle of neurochemicals'. No response is cut-and-dried, predetermined; every reaction is a subtle and complex response to the signals the brain is picking up, adjusted for inbred instincts and for the mood and feeling of the moment. The moods and feelings come from the diffusion of chemicals through billions of cells and from changing balances of positive and negative charges along the lengths of axons. The inbred instincts are reflected in the varying widths of trillions of synaptic gaps. The brain is replete with analog mechanisms: it is living flesh resonating and coexisting with the analog natural world around it.

A computer has one binary code for a yes and one for a no. A person, by contrast, may choose between 100 different ways of saying yes or no, each communicating through the nuance of voice tone a different meaning – a different level of understanding the question, of certainty about the answer, or of comfort with the context. The listener is another person, whose mind is likewise a product of the same four billion years of evolution, not of fifty years of computer science. He or she picks up every one of those nuances.

A computer can 'think' only digitally, taking decisions based on preloaded inputs and criteria. It shuffles its long streams of software backwards and forwards, replacing 0s with 1s and 1s with 0s in accordance with the rules which the human programmer wrote in. Manipulation which is logical has the merits of precision and clarity but, by the very nature of deductive reasoning, cannot have a trace of originality. Call it structured versus unstructured or computer versus human, it is the same trade-off. This is a very old theme: the logical versus the creative.

A person reaching a decision does not compute against a well-defined set of facts and criteria; instead, he or she mulls over a vast body of knowledge that is tucked away in the recesses of the mind. Even the retrieval of a simple memory is not like pulling a document out of a filing cabinet. The memory is combined with others to produce a new thought. It is a creative process. The distinguished neuroscientist, Nobel Laureate Gerald Edelman, writes of the brain that 'unlike a computer, it has no replicative memory; it is historical and value driven'.

The mind has been honed over the millions of years of its evolution to produce the balance of mental processes with which we are equipped today. It uses its spectacular processing power to produce something unique to itself: creative knowledge. It takes in images and sounds coming from the optic and auditory nerves and makes them mean *something more than just the data.* It takes electrical impulses and turns them into ideas.

About 5,000 kilometres to the east of Kerr's fictional Los Angeles Gridiron Building stand the twin towers of the World Trade Center in downtown New York City. In its 110 floors a workforce the size of the population of a small city makes electronic contact with the world outside.

As citadels of computing and communications power, these twin towers fall not far short of the fictional Gridiron. Each occupant who has a computer on his or her desk, or even just a digital cellular phone in his or her pocket, is using more processing power than did the *Apollo 11* spacecraft which put Neil Armstrong and Buzz Aldrin on the moon some three decades ago. Copper cabling runs down the elevator shafts to the basement, and these bundles of wire become as thick as a human torso as they enter the system of tunnels linking the building to the NYNEX telephone company's digital exchanges. Alongside run fibre-optic strands, each thinner than a human hair but able to transmit in one millionth of a second the digitized contents of the book you are currently holding. On the roofs of the towers, clusters of microwave dishes send data at the speed of light to orbiting satellites.

The World Trade Center does indeed epitomize the information age, but not because of the technology packed into it. It was built before the days of computers in offices, its purpose being to house the burgeoning knowledge work-force. The transformation to a 'knowledge society' took place before the emergence of personal computers and digital networks. It was and is a revolution driven by people. The signboards in the entrance lobbies of the twin towers read like rollcalls of post-industrial society: publishing firms, law practices, advertising agencies, travel bureaux, insurance companies, stockbrokers, art dealers, designers, media companies . . . And the wiring and the computers which now surround the World Trade Center's occupants are the servants, not the masters. The masters are the ideas-creators: people who, like the overwhelming majority of the readers of this book, work full time in the knowledge economy rather than on a farm or a factory floor.

The computer revolution is a subplot in a bigger revolution: the explosion of human knowledge in all its forms – progress in science and technology, innovation in business and the professions, new modes in arts and entertainment. It is part of a much broader transformation of economy and society, oriented around ideas and creativity, which had its origins in the period of the Enlightenment.

The eighteenth-century Enlightenment was, in the words of the *Encyclopaedia Britannica* 'characterized by an optimistic faith in the ability of Man to develop progressively by using reason, increasingly dominating the natural world that forms his environment'. It contrasted with the much greater reliance of previous ages on dogma passed down through generations. It brought new approaches in practical sciences and in economic thinking – Adam Smith published *The Wealth of Nations* in 1776. There was, generally, an enthusiasm to try out new ways. The most visible consequence was the birth of the Industrial Revolution in the Midlands of England in the 1770s.

Prosperity grew with mechanization, and by the end of the nineteenth century a sizeable class of people passed their lives in mental pursuits rather than physical labour. By the year 1900 there was already sufficient economic surplus to allow 18 per cent of the US workforce to be engaged full time in the creation and distribution of knowledge, having no part in the physical processes of production. That proportion continued to rise, until by about the year 1970 it reached 50 per cent: as many people worked in the nation's offices, pushing around ideas, papers, money, instructions and advertising copy, as worked in factories, farms and the rest of the 'real' economy. In 1973, Daniel Bell wrote in his *The Coming of Post-Industrial Society* that 'what matters is not raw muscle power, or energy, but information'. The pre-eminence of knowledge and ideas over physical production has now transformed the economy. For example, Coca-Cola survives by keeping its recipe private and by continually rethinking how to sell the taste as lifestyle; it is an ideas company, and has long subcontracted out the actual mixing and bottling of its drink.

Deep down, the information revolution is not about technology working with data but about people working with knowledge. It is a comprehensive shift of mankind's efforts from the creation of physical goods to the creation of ideas, images and insights – data, information and knowledge of every kind.

Interleaved with the emergence of a 'knowledge society' was the invention, step by step, of technology for augmenting our natural powers of information processing: the new creativity was directed, in part, to the production of tools for its own further development. The first step was made a century and a quarter ago, when Alexander Graham Bell invented a device to simulate the ear. A sound reaching the human eardrum passes to the cochlea, or sound chamber, making thousands of sensory hair-cells vibrate and generate electrical impulses in the auditory nerve. Bell's microphone likewise had a vibrating diaphragm and a sensory chamber which converted the mechanical vibrations into electrical signals. However, instead of travelling a few centimetres along the nerve to the brain, the signals travelled along wires to distant destinations. By 1915 they were being amplified and transmitted across and between continents. The distance barrier to oral communication had been overcome.

The next milestone was the simulation of that much richer human sensory channel, the eye, with its millions of rods and cones in the cornea which convert light signals into electrical impulses that are then transmitted to the brain via the optic nerve. In 1923 a patent was filed in the USA for an iconoscope, an early electronic television camera. A mosaic of tiny caesium/silver photoelectric cells performed the function of the cornea, and a glass lens acted as the eye's pupil; a visual image became an electronic signal. By 1936 the British Broadcasting Corporation was running the world's first public television service. The launch in the 1960s of telecommunications satellites allowed video signals to cross continents and oceans. The human eye could now see around the globe.

Finally electronic technology took on its greatest challenge – to simulate not just the sensory inputs to the brain but *what goes on inside it.* If we could copy the workings of the optic nerve as it sends images from the eye to the visual cortex, perhaps we could move on to emulate what goes on within that cortex – and then even deeper inside the folds of grey matter?

Since there are twenty billion or so neurons in the average adult brain, the notion arose among the computer fraternity that this was the number of basic logic elements, or switches, which technology should aim for to emulate a human level of intelligence. The most powerful arrays of processors in the world now contain 200 million switches. The speculation is that, by about the year 2010, there will be an installation somewhere in the world – perhaps in a military complex belonging to a superpower – that will reach the goal of matching the human neuron count.

There were those who thought in the 1970s, during the heyday of enthusiasm for artificial intelligence, that computers would provide a wholesale substitution for human mental endeavour. We would no longer have to train if we wanted to be artists, because programs could generate 1,000 images and combine them tastefully in whatever combination we desired. If we needed a business idea we could switch on the computer and programs would search for gaps in the marketplace or find money-making schemes we could get into. If we faced a health or depression problem, online advice from doctor-software or counsellor-software would be there to help us. The computer would make the old human skills irrelevant.

As recently as 1995 *Business Week* ran a feature on technology entitled 'Computers that think are almost here'. Dominating the article was a photographic image of a human brain, and it was surrounded by paragraphs headed 'Neural network', 'Genetic algorithm', 'Expert system', 'Fuzzy logic'. The subtitle of the feature spelled out, to avoid any lingering doubts, that 'the ultimate goal of artificial intelligence – human-like reasoning – is within reach'.

But researchers working in life-science laboratories despair of the crudeness of such analogies. Comparing a neuron to a single silicon switch? A neuron is a whole living cell, a hugely sophisticated processor of materials, energy and information in its own right. We are only beginning to get a feel, through the most powerful magnifying instruments, for the millions of component substructures (among them the DNA double-helix) which make up a single cell. Molecular-scale elements called microtubules are able to adopt configurations which, perhaps, provide for information storage and processing – we do not know for sure but, if they don't, other parts of the cell must. The intelligence of a single-celled organism less evolved than a neuron, such as a paramecium, is such that it can navigate towards food and negotiate obstacles, recognize danger and retreat from it. How does your PC compare?

There is a cultural divide here. Computer engineers talk of matching the power of a brain. Biologists look into their microscopes and wonder if we have matched the computational power of a single one of its cells. We don't have just the power of a single computer in our heads; the true comparison would be with a figure more like *twenty billion computers*. The complexities involved are genuinely difficult to imagine. If your brain-cells' DNA strands alone were unravelled and placed end to end they would stretch for over 150 million kilometres – from where you are sitting to the sun, through it and beyond.

If you could tour the twin towers of the World Trade Center when they were deserted of human life, in the early hours of the morning, you would be able to count a million dollars' worth – more or less – of digital electronic equipment on each floor. But if you chanced upon a security guard you would confront, in that single person, more processing power than is present in the electronics wired into all of those 110 storeys.

★

There is a fight on now. Not a physical fight between computers and people, as in the ISAAC story, but something more subtle: a battle for hearts and minds in the digital age. It concerns how much we should be prepared to adapt our lives in response to the new technologies. There are those who see the computer ushering in a new phase of civilization, in which our aspirations, the things which are important in our lives, change in the light of the capabilities of digital computers. They place great hope in the ability of technology to address the problems we face in our lives. Books like *The Road Ahead* (1995) by Bill Gates herald a future in which computers and networks cater to ever more of our needs and desires. Computers will play their part alongside us in thousands of little ways, like putting on music they know we like as we enter the house, as well as in more significant ways, like replacing fallible human medical care with superlative software-based diagnoses. The futurologist Alvin Toffler believes that the information revolution will change our lives more than the combination of mankind's two previous revolutions – the transition from a hunter-gatherer society to an agricultural one and the subsequent transition from the agricultural to the industrial era. In this vision, the Information Superhighway will replace commuting and shopping; new generations of computers will make much human mental endeavour redundant; cities will be emasculated; we will live and work in cyberspace.

For our work, our families and our children, the increased influence of computers is seen as not only inevitable but good. In California, spiritual capital of the new technology, a taskforce of forty-six teachers, parents, business executives and technology experts in the summer of 1996 urged the state to spend $11 billion on computers, concluding that this technology 'more than any other measure . . . offered the potential to right what is wrong with our public schools';

among the alternatives their report had considered were reducing class size, improving teachers' salaries and facilities, and increasing hours of instruction. This reflects an outlook on the world that is rapidly gaining currency: that computers and networks are the defining characteristic of our time, the most dramatic of the forces for change affecting society. We must get accustomed, in the words of MIT's Nicholas Negroponte, to 'Being Digital'.

There is, however, an alternative view. Put the forty years of computing achievement into a longer and broader perspective and a different story emerges.

The predictions that computers and networks will deliver a wholesale basket of substitutes for our traditional needs and desires presuppose a view of human nature which not everybody shares. In January 1998 Howard Segal, of the University of Maine, wrote in *Nature*:

> The transformations these technological elites deem almost inevitable are surely far from certain . . . the dreams of these engineers and inventors are more pathetically naïve than enviably sophisticated.

Even with the development of virtual-reality systems, the notion of a large-scale substitution for live contact with real people and tangible goods is to underestimate greatly the subtlety and sensitivity of the human sensory processes. Writing in 1997, Michael Dertouzos, head of MIT's Laboratory of Computer Science, reminds us that the digital product will always be, at some visceral level, 'unbearably fake' – that we are still biologically ancient humans and that the primal forces of the cave which really move us, like friendship and fear, do not transmit through networks. 'Deep down,' he comments, 'our psyches know that 1s and 0s cannot love, nurture, hurt, or kill us at a distance.'

The human animal, owner of that most complex of machines, the brain, has developed intricate fabrics of social and economic institutions over thousands of years. It has built these institutions in its own image; it has not modelled them on machines. Within these fabrics, each of us pursues our dreams and ambitions. The pursuit is not computer-like. In the context of this social and economic fabric the great strengths of computers – that they are precise, crisp and logical (in short, digital) – are also their great weakness. It is the unfathomed minds of people, with all their logical failings but also with all their creative energy and original ideas, which will continue to be the driving forces in business, politics, art and society.

We cannot know where the explosion of ideas and knowledge which we call the information age, with its remarkable new technological toolset, will lead us. The record of prediction with regard to information technology is very poor. Professional forecasters – academics, industry analysts, futurists – have tripped at every step. In 1943 IBM Chairman Thomas Watson forecast that there would be worldwide demand for just five computers, none of them in business; by the 1960s there was at least one computer in every large corporation. In the 1970s the industry and its advisers failed to predict the personal computer; as late as 1977 Ken Olson, president of Digital, said: 'There is no reason for any individual to have a computer in their home.' It was left to two young students working in a California garage, Steve Jobs and Stephen Wozniak, to create the machine which would bring the digital revolution to the masses. (Even science-fiction writers, not usually lacking in technological imagination, did not foresee today's home computers.) In the early 1980s the pundits had not yet come up with the word 'Internet'; by the early 1990s they were talking of little else. And these are the experts, the ones who (to borrow ironic verse from John Betjeman) 'really ought to know, for they are paid for saying so'.

What is least likely to change is who we are. The pace of natural evolution is exceedingly slow compared with that of technical progress. Science produces surprises every year, but our eyes, ears and brains develop only on evolutionary timescales, typically requiring tens of millennia. The human brain has a certain capacity to take in and assimilate information, to think it through and act wisely or unwisely upon it. For all practical purposes we can assume that the human being – a few more or less outrageous experimental subjects apart – will be unchanged as a biological entity half a century from now.

This book tells of the emergence of a remarkable new technology, the technology of information processing. From prehistory until the dawn of the electronic age the ability to manipulate symbols, images and ideas was limited by the power of the human brain. Less than fifty years ago, scientists began to produce silicon devices which had the capacity to process data electronically. From these inauspicious beginnings there emerged an industry which has overtaken all other industries in the race for resources and markets. Now computers magnify the reach of our senses, as the machines of the industrial age magnified our muscle power. We can create a virtual world which seems to have as much richness and value as the tangible world from which it sprang, and is in its own terms better: the images are sharper and the logic more precise, so that we seem closer now to the ideal world of Forms conjured up by Plato than we have been for 2,000 years.

In the first thrill of discovery, cyberspace seemed to provide quick solutions. The new revolution allowed us to travel the networks – to glean information from every corner of the earth. Millions of hopeful travellers sat at their screens looking for answers they had not found in their own milieux. Like nineteenth-century visitors to unexplored African lands, we marvelled at how much was out there.

But your mind or mine will make use of only as much information as it can absorb. Those of us living in twentieth-century towns and with access to television channels, telephones, bookshops and newspapers already have on tap more information than we can take in. The problems which matter most in our lives will not be answered as if by magic on our computer screens when we buy a modem. We will, as much as before, have to build the means to resolve them inside our heads.

And we certainly do not need to buy into a new philosophy of life, a sort of cyber-ontology in which the meaning of existence has been solved by deciding that we are computers. Our minds don't just shuffle bits to and fro, rearranging them according to preset rules. The idea that something transcending the logical manipulation of data is going on inside our minds is not obscurantism: there is a genuine mystery here which the new technology has highlighted, not solved. Professor Richard Dawkins of Oxford University, drawing analogies between the genetic code in our cells and computer data, wrote in 1995 that 'this digital revolution has dealt the final, killing blow to the belief that living material is deeply distinct from nonliving material'. We can respond, like Mark Twain to the premature publication of his obituary, that reports of our demise have been greatly exaggerated.

But that is to jump ahead. We must first go back, to a time long before the age of technology.

2

The Ancient Mystery of Human Knowledge

The year is 8000 BC. A hunter-gatherer is standing amid the plains of Northern Europe, scanning the horizon for animals – animals which might be food for him, or which might represent a danger.

Although he wouldn't think of it in these terms, the panorama before him is generating immense numbers of photons. Every leaf, stone and passing bird is sending out beams of reflected light at varying levels of frequency and intensity in accordance with the source's shape, colour and texture. These streams of photons are data – raw signals – and they number in the millions upon millions. The pattern of leaves on the trees alone contains more data than all the pages of the *Encyclopaedia Britannica*.

But as the light rays pass through the air they are not yet information. Nobody has yet seen them; nobody has yet been *informed*.

Some of these photons reach the eyes of the hunter. They are focused onto the rods and cones of his retinae, causing signals to be sent via the optic nerves to the visual cortex in the brain. A great deal of processing happens along this route. What his conscious mind registers is not raw data, but a distillation of it which he can digest and make sense of. He sees patterns and regularities in the light signals, and these he interprets as trees, rocks and birds. Only now has the data become information.

Information is data interpreted by the person who is being informed – in this case, the hunter-gatherer. Since information is a distillation of the data, its quantity is more limited. From a million photon streams our hunter-gatherer identifies perhaps 100 distinct objects: a million streams of data become 100 items of information.

He now knows something more about the world than he did a few moments ago, specifically about the scene he is observing; he knows, shall we say, that the river is unusually high for the time of year. The information has altered his state of knowledge. Knowledge is accumulated information in the brain. But it is not simply the sum of what went in, in the way that a library is the accumulation of its books. It is the result of a process that goes on in a highly evolved part of the brain's cerebral cortex, the tertiary region. Just as information is data distilled and interpreted, knowledge is information distilled and interpreted. From the 100 objects our hunter-gatherer sees as he surveys the scene, he gleans just a few new insights.

In short, with this brief scan of the horizon the hunter-gatherer's eyes receive *data*; his optic nerves and visual cortex process it into *information*; his brain's cerebrum makes it into *knowledge*.

Information has shot to prominence in the electronic age. But it is not a creation of the new technologies. Humans have always been outstanding processors of information and of its close cousins, knowledge and data.

We use the word 'information' very freely in countless contexts: information processing, information overload, insider information. But what exactly do we mean when we refer to this entity that has no size or weight but which can undoubtedly have value and cost? Through an accident of our intellectual training, we are curiously ill-equipped for discussion of such a question. Although the events of the last four decades have catapulted information to a central role in economic life, even well-educated adults generally cannot produce a cogent definition of this somewhat abstract concept. And they have even more difficulty discussing it quantitatively. What is the information content of a word, an image, a DNA strand or a snowflake? Ironically, though we live at the turn of the twenty-first century, 200 years after the agricultural age began to be superseded by the industrial, we are still more comfortable with agricultural measures than with information ones. A group of computer-based traders on Wall Street, taking their sandwich break, could more easily discuss the relative yield of land sowed to wheat versus barley than assess the amount of information in a drawing or a piece of music. They would be familiar with kilograms per hectare and other basic units of agricultural output, but not with entropy as a basic measure of information.

'Where is the wisdom we have lost in knowledge? Where is the knowledge we have lost in information?' asked T. S. Eliot in *The Rock* (1934). The word 'information' suggests a flow of facts, the act of informing. Knowledge implies something deeper and more lasting, a body of things known. Wisdom connotes still greater depth of meaning, knowledge truly understood. In modern usage we have another concept, data – just raw symbols out of context, not interpreted. Were he alive today Eliot might add, 'Where is the information we have lost in data?'

What really constitutes knowledge has been a central puzzle of philosophy, more precisely of epistemology, since antiquity. Socrates was convinced that he knew nothing; on being told by the Delphic Oracle that he was the wisest man in Athens, he asked all the reputedly well-informed men in the city how knowledgeable they thought they were. On hearing their overconfident replies he concluded that the Oracle was right, since he at least knew that he knew nothing!

René Descartes, the first philosopher of the modern period, was prepared to go further. When he wrote in the 1640s 'I think, therefore I am' (surely philosophy's best-known quote – he had an early sense of the power of the sound-bite), he was stating the one thing about which he was prepared to be absolutely certain. In the eighteenth century Immanuel Kant, acknowledged as the greatest of the modern philosophers, reflected long on how information coming through the eyes, ears and other senses could become meaningful knowledge in the mind. He concluded that the mind had to incorporate a structure of 'prior' knowledge, which allowed incoming sensory knowledge to be understood. Without that structure in the mind there would be no possibility of gaining knowledge of the world through observation; the incoming information would be just a collection of meaningless symbols. The depth of Kant's analysis was unique, but the notion of knowledge being divided into two kinds – prior or innate ideas present in the mind and experiential information coming in through the senses – has been prominent in Western philosophy through the ages.

Kant studied the mind through contemplation. We can now look at the brain through powerful electron microscopes. The physiological evidence suggests that, as Kant predicted, some knowledge is innate and some acquired through the senses. By the moment of birth every brain-cell which a child will ever have is in place. The brain's enormous complexity is built up under the instruction of DNA molecules contained in the single cell which is a fertilized egg, and which is faithfully reproduced in every cell in the body during growth, to be passed on (in conjunction with DNA from a person of the opposite sex) to the next generation. The DNA molecule, a device that has evolved over four billion years – the history of life on earth – is the mechanism whereby each human being transmits to the next generation the secret of the construction of his or her body, including a goodly dose of inbuilt knowledge in the brain. Professor Colin Blakemore of Oxford University has described this as our 'inherent wisdom, the distant memory of the experiences of ancestors'. DNA is, biologically, the source of what Kant called prior knowledge.

The obtaining of knowledge by making inferences from the factual data we pick up through our senses is inductive logic. Its most ardent critic was the eighteenth-century Scottish philosopher and historian David Hume, who was a friend and colleague of the economist Adam Smith. He refused to accept the validity of any inference made solely on the basis of empirical observation. Even though night has always followed day throughout the memory of mankind, he famously rejected as unproved the general proposition that night follows day; he regarded as still open to doubt the conclusion that *today*'s day will also be followed by a night. He could accept the validity of the logic of deduction, by which specific facts are deduced on the basis of a generalization, but not that of induction, whereby a generalization is inferred from specific facts. Hume's scepticism, however, seems to have lost out over time to the empiricism of his near contemporary John Locke, and of countless subsequent philosophers and scientists: the inference of knowledge from observed facts is the basis of all modern science.

After the moment of birth our base of knowledge is increased greatly by information coming in through the sense organs. At any one time each of us has a given set of beliefs or ideas, held with different levels of certainty, which constitutes our state of knowledge. By further observation of the world that state of knowledge is changed – it is developed or refined. In every one of our waking seconds, megabits of data are reaching our eyes and kilobits of data are reaching our ears.

Data – the word is Latin for 'things given' – are raw physical signals or symbols. The data inputs to the human body are the five familiar senses – sight, hearing, touch, taste and smell – plus a few others we use as often but easily forget, like the senses relating to skin temperature and balance. Animals vary hugely in their reliance on different senses. In humans the richest sense organs by far, in terms of the volume of data processed, are the eyes and ears – in that order.

The eyes and ears have pride of place because they work by detecting oscillations in the environment – respectively, light and sound signals. Oscillations – waves, cycles, frequencies, rhythms – are at the heart of any good communication system. Facts can be communicated by other means (an odour wafting through the atmosphere) but a wave – an oscillation of light, electricity or sound – has the advantage that it propagates particularly quickly and evenly, without losing its key identifying characteristic: its frequency. The eyes and ears pick out, with remarkable precision, the different frequencies present in light, perceiving them as different colours, and in sound, perceiving them as different pitches.

When rays of light from a scene reach our eye, the natural lens in the cornea, helped by a spectacle lens if we are short- or long-sighted, focuses them onto the retina. A tiny image of the scene is created on the receiving sensors in the human eye, the retina's rods and cones. There are over 100 million rods, each sufficiently sensitive to be triggered from an 'off' to an 'on' state by a single photon of light. The ten million cones are more sophisticated, although they require higher light intensities to be activated: they can identify colour, some responding to the frequency of blue light, some to yellow-green, and the rest to red.

The receiving sensors in the ear are tiny hairlike structures, specialized nerve-cells which pick up waves of pressure in the air (sound waves are compressional, unlike light waves, which are electromagnetic oscillations). Listen to a clear note played on a musical instrument and the hairs tuned to that frequency will resonate and trigger neural activity. Listen to someone talk, and hairs of many different tunings will resonate, because the human voice is a melange of numerous frequencies. Then there are receptors placed all over the surface of the body that detect heat, cold, pressure and pain, each able to send signals along a nerve-cell to the brain to register the sensation.

The quantity of signals being generated by the natural world in any second of the day is almost limitless, far greater than any receiving sensors, human or technological, could assimilate. The retina in the eye of the hunter-gatherer scanning the horizon registers incoming visual data as millions of streams of incoming photons. The data are light rays, objective signals, things given. Information (from the Latin *informare*, meaning 'to describe') is an abstraction, a description or statement. What gives the signals significance, what makes them represent information, is the way they are then processed.

The processing of streams of data into meaningful information begins within the eye, which decides what is useful even as the data strikes the retina. The first priority is to detect movement, which could be life-threatening – it might indicate a predator closing in. Waving your arm energetically to catch the eye of a taxi driver may be inelegant but it is the biologically correct action; if the driver is looking, as he should, straight ahead, your moving arm represents your only hope of triggering a response from his rods and cones.

The next priority is to favour the centre of the image. If you need to look at something carefully you instinctively look straight at it: the middle of the retina is where the sensors are packed in tightest. This prioritization is more extreme than you might think, for you actually see surprisingly little of what is in front of you: the part of the retina which can see even moderately sharply covers less than 1 per cent of the field of view, an angle so small that without moving the eyes you cannot read both ends of a longish word. (Fix your sight absolutely rigidly on the *y* in *absolutely* and you'll find that the front half of the word is a jumble of unreadable letters.) Apart from a tiny sharp spot dead ahead, our eyes pick out only rough patches of light, like the brushstrokes in a French Impressionist painting.

By the time the signal has been processed – or preprocessed, we might say – within the eye it has already been compressed very usefully. The optic nerve contains 'only' one million transmitting channels; over 100 million signals from the primary sensors have therefore already been compressed one hundredfold before the information leaves the retina. This greatly reduces the demands on the next stage of processing but, even so, the upcoming task is daunting.

The way the brain processes visual data (the million signals in the optic nerve of the hunter-gatherer) into information (*this is an antelope*) is a masterpiece of Nature. Computers have great difficulty with this task. Vision systems – computer systems which can analyse light patterns and recognize objects – are still in their infancy: the humble office coffee-cup baffles even the most sophisticated program when it is placed against a new background, or merely turned upside down. But we should not be harsh on the programmers. It often seems that an infant in a cot is spending much of its young life staring endlessly at the objects around it. What is happening is that its brain is hard at work getting good at analysing visual data. Of its (roughly) twenty billion neurons, a substantial proportion is in the visual cortex, specialized to the interpretation of signals coming in from the optic nerve. Many months of staring, the occasional prod, and the infant will finally be able to recognize a cup – and other household objects. It will take a while before software can emulate this feat.

When our primary cortex regions (visual, auditory, olfactory, and so forth) preprocess data for passing on as information to the secondary and tertiary ('thinking') regions, they do so in ways optimized – we must presume – for that subsequent thinking process. Nature has taken four billion years to evolve the human brain, and the processing of the signals produced by eye, ear and nose to the purpose of subsequent mental analysis is supremely subtle; computer scientists will not reproduce it with a few years of programming effort.

So far, the hunter-gatherer has processed the rays of light (data) reaching his eyes into images of animals moving in front of him (information). That is the easier bit. It can be done by a two-year-old child, or even by a very sophisticated electronic vision system.

The next stage is for him to work out what the images *mean*. For this he puts to use not just the visual cortex but the thinking parts of the brain – the secondary and tertiary regions. He has to delve deep into the recesses of his knowledge base, to rely on his experience of Nature, to use his intuition. The distillation of information into knowledge is not the same as interpreting a stream of photons to be a moving object: it is the result of an enormously complex application of previously learned facts and ideas. It is a higher function of the brain, carried out in the tertiary region (the one most adapted to abstract ideas) of the cerebrum. We call it, quite simply, thinking. The act of thinking relies not just on the incoming visual data but on a vast reservoir of pre-existing knowledge residing in the mind, both inherited from our ancestors and based upon the experiences of our own lives to date.

Knowledge builds up over time. Within an individual it builds up during the course of a lifetime, the accumulation being countered to some extent by loss of memory. It also builds up to a much greater extent in the human race as a whole. Although the marks on the paper in books and scientific journals cannot be considered to be themselves knowledge, they do encode the knowledge of previous generations and allow the current generation to build up its knowledge base much more rapidly than would otherwise be the case. Trying to cover the full range of modern biology through books may seem irksome to the frustrated undergraduate, but it is certainly more efficient than trying to repeat all the original research oneself.

The use of the terms 'knowledge' to refer to facts which have a lasting value and 'information' to refer to facts of more transient relevance is consistent with everyday English usage. This distinction is consistent with the etymology of the words 'knowledge' and 'information'. The root of the word 'knowledge' is the verb 'to know', which refers to a state of knowing, a continuous or ongoing state. The word 'information' stems from the verb 'to inform', the act of communicating facts rather than the state of knowing them.

We would say that an encyclopedia or a volume of the *Journal of the American Academy of Sciences* contained knowledge, and that an airline timetable or a telephone directory contained information. Knowledge has lasting value; information is

Early analog: an image of a bison drawn on the walls of the caves at Lascaux, perhaps 30,000 years ago. Today, the power of the television image testifies to the importance of the analog method – of communicating by producing an analogy, a semblance or likeness of the original.

valuable for a certain period of relevance, after which it is superseded – by a new timetable or a new directory. The *receipt* of information can be of permanent value, but only when the information adds to that body of more lastingly useful facts which is our knowledge. A chemistry experiment conducted by a school student as an exercise in class yields information; an original experiment performed to current scientific standards in a research laboratory may produce information that adds to the body of our knowledge.

One of the great triumphs of humankind has been to develop means of recording knowledge for future generations, so that the sum of human knowledge can accumulate through time on a grander scale than could be achieved by any individual parents teaching their own offspring.

The caves at Lascaux show that humans were drawing representations of bison on walls at least 30,000 years ago.

Signifying an animal by drawing a scale representation – the analog method – survives as art, but as the norm for functional communication it has been replaced by alphabetic text. Instead of drawing a bison (carefully, so that it is not mistaken for a boar or a buffalo) we just write the five letters b, i, s, o and n. The new form achieves the required representation much more quickly (time yourself writing those five letters against drawing a picture of a bison) and, to those who have learnt the language, does so unambiguously.

The first written alphabet was invented in the Eastern Mediterranean between about 1700 and 1500 BC. Alphabetic language marked a great step forward in human information processing: it was digital. Alphabetic writing is one of the two great triumphs of the digital encoding of information; the other, even more obviously, is numbering. As any digital coding system, alphabetic writing is based on a manageable number of discrete symbols. In the case of English, there are twenty-six of these:

a b c d e f g h i j k l m n o p q r s t u v w x y z

These twenty-six basic elements combine in a vast number of combinations to provide the words on these pages and on the untold millions of pages of other books.

The art of writing began much earlier, being developed in Egypt around 4000 BC and in Babylon shortly thereafter. The oldest writing systems, however, were not alphabet-based. They began as pictures of the objects being described (pictograms), and then the pictures became standardized into ideograms (such as Egyptian hieroglyphs), each representing a different item or concept.

Ideogram-based writing systems have great power and beauty, and survive to this day, most notably in China and Japan. But if viewed from the standpoint of modern information theory, they are less efficient than alphabets as coding systems. They contain an enormous number of characters, instead of taking advantage of the economy of representation afforded by using a small number of digital elements (letters) in different combinations. This makes them very difficult to learn – Japanese children spend a large proportion of their schooling until the age of 13 or 14 simply learning to read and write their own language. It also makes them more difficult to use in an age of digital technology.

The letters of the alphabet are entirely artificial constructs of human civilization. There is nothing about the shape of the letters 'd', 'o' and 'g' to suggest the shape of a dog. The symbols were created artificially, and have to be memorized 'artificially' by every succeeding generation of children. In this respect writing differs from the ability to produce

Early digital: a fragment of Greek text from the Rosetta Stone. In digital communication, information is encoded using an agreed set of symbols, such as numerals and letters.

spoken language, which appears to be 'natural', pre-existing in the brain. In the days of nineteenth-century colonial expansion European explorers discovered thousands of human tribes, every single one of which had a well-developed spoken language – at least 8,000 spoken languages were counted (there are many fewer now) – yet the overwhelming majority of these peoples did not write. Just how deeply innate are language structures we do not yet know; the subject is fiercely debated. But there is no doubt that, while humans have an instinct to *speak*, they do not have an instinct to *write*. The digital structure of written language may have become central to advanced societies, but it is not actually coded into our nature.

The other great digital invention is, as we have noted, the system of numbers. The earliest numerals of which we have record appeared in Egypt around 3400 BC; they consisted primarily, and not surprisingly, of simple straight marks for the small numbers, but there was also a special form for 10. We can thus see that decimal coding, which survives to this day as the main system for counting, caught on early; 5,500 years ago is only about half as distant in time as the end of the hunter-gatherer (or pre-agrarian) period of human prehistory. By 3000 BC the skill of the Egyptians in mathematics, and in writing, was sufficient to support a class of professionals, the scribes. The scribes performed duties similar to those nowadays performed by the civil service, including tax accounting and collecting.

The idea that quantities – amounts, sizes, weights, times – are recorded digitally is today so familiar that it takes some effort to recognize that in Nature quantities occur in analog form, and that the way we represent them is, as with alphabetic language, by means of an invented code. To particularize, we have two types of wristwatch, analog and digital. The clockface of an analog watch is a picture of the day, and the passing of the hands around the face represents the passing of time. In the earliest clocks, sundials, the analogy was self-evident. In a modern watch it is more intricate, for the hour-hand goes round twice for every one rotation of the earth, the minute hand twenty-four times. But the defining element of analog coding is there: a representation which is analogous to the real thing, in some sense a facsimile.

A digital watch does not present a visual representation of passing time, but a number – for example, *10.37*. The shape of the number tells you nothing about where you have reached in the course of the day. As with letters of the alphabet, the meanings of numerals have been arbitrarily defined by their human inventors: the symbols are pure code. There is nothing about the appearance of the numeral 7 which indicates that it represents a larger quantity than the numeral 5.

In the contrast between the two types of watch you see much of what is good and bad in digital and analog. The digital has, to its credit, precision: *9.58* means exactly what it says, and does not mean *nearly ten*. This precision aids subsequent processing of the data; if you want to know how many hours there are between two and nine o'clock you will do the sum digitally (nine minus two equals seven), not measure it out with a ruler around the watchface. And for recording or transmitting to others, the digital system is more compact: bus timetables are full of digital numbers, not little pictures of clocks showing where the hands will be when the bus arrives.

The benefits of analog are more subtle. Because the hands present a picture of the time, you read the meaning spontaneously, without the small but noticeable intellectual effort of decoding a set of digits. While from a display that says *7.43* you can calculate quickly enough that it is time to run for your ten-to-eight train, from the analog watch you actually *see*, directly, how fine you are cutting it – because the two times are so close together on the dial. Everything about the analog system is a little more natural. The hands appear to move smoothly, like time itself does, not incrementally, in the manner of an LCD display. While these benefits of analog watches may be subtle, they are real enough to sway purchasers: sales of digital watches, even though they are now cheaper, still lag behind those of analog ones.

As a consequence of the invention of numbering and the alphabet by Mediterranean peoples some 4,000 years ago, the great majority of human recorded thought is now encoded in discrete symbolic (that is, digital) formats using 10 numbers, 26 (or thereabouts) letters and a few punctuation marks. This fact has been of enormous practical consequence for the development of information technology. There are 24 million volumes in the US Library of Congress, and three and a half kilometres of new books arrive each year. Every man-made fact or proposition, from the US Constitution to the principles of operating a petrochemical complex, has been encoded into these 10 + 26 symbols, ready for manipulation in accordance with the rules of logic. The conversion of the letters and the numbers into binary form, by encoding each of the 10 + 26 characters in turn in an even more readily manipulable form, was trivial by comparison. Each character simply had to be given a several-digit binary equivalent.

It is humbling to reflect that the key step towards a technology of information was taken 3,700 years ago when Mediterranean peoples alphabetized (i.e., digitized) written language. Had this not happened, twentieth-century computing technology would have had to work directly on the natural – analog – sounds of human language. Even today, with gigabit-per-second parallel processing available, the interpretation of speech by computers is a technological nightmare, cumbersome and unreliable. The digital revolution has benefited beyond measure from the uniformity and precision that was introduced into human communication when spoken language was converted into the alpha-numerical symbols of the written form.

This achievement of the ancients was to receive its grandest embodiment in the library at Alexandria. The city, founded by Alexander the Great in 332 BC, became within a century the intellectual capital of the world and a great centre of scholarship and science. By the time of the goings on of its famous residents Antony and Cleopatra in the first century BC, its library had grown to hold hundreds of thousands of books, and had lost many of them to the first of its several war-inflicted tragedies. Cleopatra was in fact greatly attached to the collection and is credited with rebuilding much of it, helped by the spoils of Antony's military conquests. The repository of knowledge continued to grow in breadth and scale for four more centuries.

The library at Alexandria was sacked and burned in the year AD 389 by a Christian mob energetically following the order of Emperor Theodosius that all pagan works should be destroyed. A small part of the collection remained intact until the library suffered its second, and definitive, burning at the hands of Muslim attackers in AD 642. This latter act of destruction has come to symbolize the beginning of the Dark Ages. In those days it was rare for there to be more than a single copy of a book in existence: through these two events, therefore, a great part of the accumulated knowledge of the ancient world, including much of its science and most of its mathematics, disappeared. Much of it would not reappear for over 1,000 years.

The city never regained its former pre-eminence – by the time Napoleon invaded Egypt in 1798 it had become a small fishing village. But there will be at least a symbolic marker to remind us where the world's greatest base of knowledge once resided. In 1987 UNESCO and the Egyptian Government backed a project to build and stock a new great library in the now sprawling modern town. Sited on the presumed location of the original library, in the ancient royal quarter, this monument was scheduled for completion, at a cost of $170 million, in time for the millennium.

Having developed digital means for communicating human ideas – characters and numerals – the next step was to formulate digital rules for manipulating them. These are the rules of logic. Logic, with its sister discipline mathematics, operates at the second level of the communication–processing–thinking hierarchy: processing.

The idea that you can engage in crisply logical reasoning of a yes–no character – if Mr A is older than Mr B then Mr B is not older than Mr A – must have occurred to humans in prehistoric times, but the production of systematic principles for such reasoning is attributable, as is so much else in our intellectual heritage, to the Greeks.

The sixth-century BC polymath Pythagoras of Samos was a mysterious figure who achieved such cult status during his life that biographers find it difficult to separate fact from myth. He was certainly a hugely influential figure: the school of philosophical thought which he founded remained vibrant, in various reincarnations, until well into the Middle Ages. On the basis of a belief in the transmigration of souls, he set out principles for a way of life which were given concrete manifestation in the form of the Pythagorean Brotherhood, a monastic-style order to which aspirants could apply. The successful contenders had to accept many quasi-religious tenets: they had to wear only white clothes, they had to observe sexual purity, and, more oddly, they had to abstain from eating beans. Above all, however, they had to demonstrate exceptional skills in mathematics.

Pythagoras rightfully occupies a special place in the history of mathematics, which he elevated from a tool for achieving practical goals to a status of its own. The principles he formulated in the century before the time of Plato and Aristotle contributed much to the development of Western rationalist thinking. He saw in mathematical logic a purity of thinking worthy of the highest respect – indeed, of worship. Even in our scientific age the metaphysical status he gave to numbers – the concept that reality (as manifested in, for example, music and astronomy) was at its deepest level mathematical – does not seem out of place. The statement by early Pythagorean philosophers that 'all things are number' resonates particularly well in the age of binary-coded virtual reality.

Earlier civilizations had become very adept at arithmetical operations, and Pythagoras studied the techniques of the Egyptians and Babylonians. What he developed was the concept of numerical logic, of proof. Most famous of the theorems deduced was the one which states that in a right-angled triangle the square on the hypotenuse is equal to the sum of the squares on the other two sides. Junior high-school stuff now, but a huge achievement at the time, and the Pythagoreans recognized this; it is recorded that 100 oxen (no slaves or virgins – the Brotherhood was exotic but humane) were sacrificed to thank the gods for this insight. What was rightly judged momentous was not that triangles had this feature (which had long been suspected) but that one could deduce this conclusion logically and infallibly. Simon Singh writes in his book *Fermat's Last Theorem* (1997):

> What seems certain is that Pythagoras developed the idea of mathematical logic . . . He realized that numbers exist independently of the tangible world and therefore their study was untainted by inaccuracies of perception. This meant he could discover truths which were independent

> of opinion or prejudice and which were more absolute than any previous knowledge.

If triangles, what else? The power of logic came to fascinate the Greeks. For some, like Plato, the certainty of mathematics was held as a model for reasoning in other spheres of life, including ethics and politics. To this day logic is held up as the gold standard of human reasoning.

A standard, yes, but one that we cannot apply to the great majority of the choices which we make daily. In ethics and politics, and in most spheres of life, we mull over, think through, contemplate, act on impulse, and so forth. Plato may have wished us to live by rational logic, but history has shown that that is not the way humans behave. The great majority of our manipulation of ideas is still done according to the fuzzy way in which our minds naturally work. The digital recording and transmission of ideas via alphanumeric characters has caught on fairly comprehensively, but the logical processing of ideas only very partially. So, when the electronic age arrived in the nineteenth and twentieth centuries, the communication of ideas was ripe for automation but the processing of them much less so.

To the Brotherhood, numbers and other digital symbols represented purity. They also gave information a quantitative measure.

The analog data of the natural world – sound waves, light waves – are often measured by the frequency of the transmitted signals. The frequencies of audible sounds range from 20 to 20,000Hz (cycles per second); those of visible light from 750 trillion Hz to more than double that. These are of course measures merely of the medium, not of the message; they do not get at the content. With digitally encoded information we can count the content: characters in a printed book, bytes in digital computer memory.

The crudest data representation is the binary: every fact to be represented must be distilled to a *yes* or a *no*, a 1 or a 0. The corresponding measure is the Binary Information uniT, or bit. A *yes* or a *no* is a bit. A mixture of 20 *yes*ses and *no*es represents 20 bits – which actually allows over a million combinations. Digital information technology usually works internally with bits – two voltage levels, representing 0 and 1 respectively. Fortunately the interface with the operators is generally via a more user-friendly information unit, the byte. A byte is the name given to a block of eight bits, which allows for 256 combinations – enough to encode the 26 letters of the alphabet (in both upper and lower case), the numerals 0 through 9, and a fair number of punctuation marks and machine instructions.

The fact that a byte encodes a single keyboard character also makes it a convenient measure for text information; each written letter is a byte (eight bits) of information. On average a word in the English language has five letters, and hence represents five bytes or forty bits. A page of printed text, as in this book, may contain 500 words and hence represent 2.5 kilobytes. So a volume of a few hundred pages contains about a megabyte of text data. For those who want to have a feel for data measures in everyday terms: you are holding about a megabyte in your hands.

Actually, these are not true information measures. They are simply symbol counts, and do not permit any assessment of the extent to which the recipient is being informed. Information theory developed gradually during the course of the early twentieth century, having as its objective the development of something better than a symbol count: a measure of information which would correspond to the actual content or contribution to knowledge contained in information. The definitive work in this area, which brought information theory to its modern form, was undertaken in the 1940s by the mathematician Claude

Shannon and by Norbert Wiener, the 'father' of the field of cybernetics.

The challenge of information science was to develop a measure of information which was independent of the mode of representation – that is, which did not vary according to the many arbitrary choices about how the information could be represented. Bits and bytes are satisfactory symbol counts, but their inadequacy as information measures can be illustrated by a trivial example. If somebody wishes to know what page of this book you are reading and you pass that fact on to them, how much information have you imparted? Your response to their question, 'fifty-six', comprises nine characters – nine bytes or seventy-two bits. But what if you wrote the answer as '56'? This has just two characters (two bytes, sixteen bits) but exactly the same amount of information, just differently represented. The problem is further exacerbated if we consider different languages. What if, instead of using the English 'fifty-six' you gave the answer in German, '*sechs und funfzig*'? Here you have fifteen characters (fifteen bytes, over a hundred bits). Surely you have not conveyed more information just because you used a different language?

Part of the problem lies in the redundancy of much data as a means of representing information. The fact that the numerals '5' and '6' are a much more compact representation than 'fifty-six' implies that the full text version of the numbers includes a fair number of redundant characters – characters which could be omitted without impairing our ability to communicate the necessary information. Redundancy is a pervasive feature of most human and machine communication.

Is there not a consistent measure of the amount of information required to identify a page in this book? There is, and the measure is entropy. The word 'entropy' has other important meanings: in this context, however, it defines information not as a number of data bits but as a change in a state of knowledge. Prior to receiving the communication from me, the enquirer did not know which particular one of roughly 300 pages of this book were being read. This *uncertainty* defines the enquirer's prior level of knowledge (or, to be more precise, lack of knowledge) of the situation. After receiving your response, the questioner knows the answer, 56, with *certainty*.

There has long been an association between the concepts of knowledge and certainty. Ideally, knowledge is something about which one could be certain, as Descartes was of his existence. Beliefs held as mere *probabilities* are poorer states of knowledge. In today's mathematical theories of information, certainty and uncertainty are the measures of knowledge. Lack of knowledge is measured, quantitatively, as an extent of uncertainty; in this example the extent of uncertainty is one part in 300. After receipt of your message the questioner's uncertainty is reduced to zero: the page number is correctly known. The information content, or entropy, is equal to the extent to which uncertainty has been reduced. By convention, the unit with which entropy is measured is the bit, which is, perhaps confusingly, the same unit as that used for counting binary data digits. When the number is expressed as 300 it contains three digits but, when coded into binary, it has eight (to be specific, it is 100101100); therefore the extent of the questioner's prior uncertainty was eight binary units. The quantity of information contained in the statement that you are reading this particular page is, therefore, eight bits.

Entropy analysis shows that English and other natural languages use symbols very wastefully; that is to say, they are highly redundant. That savings can be made in the number of characters used without losing meaning is demonstrated by a traditional British classified-advertising publication, *Dalton's Weekly*, which saves space (and advertisers' money) by leaving the vowels out of words: *Sntncs r prntd lk ths.*

Greater savings still could be made, though (as in this example), at the expense of the beauty of natural language. Claude Shannon, the founder of modern information theory, calculated that the entropy, or true information content, of English text is in the region of one bit per letter, as opposed to the eight or so bits we use to write it using the twenty-six letters of the alphabet and our traditional rules of spelling and punc-tuation. However, though Shakespeare's works could theoretically be represented using an eighth of the space taken when written conventionally, the result would be, while efficient, not exactly poetry. Redundancy has its value, too.

In the 1960s, with the advent of bank computerization, account numbers were introduced to replace names and addresses of customers as the means of identifying accounts. In protest, a customer in Maine signed a Christmas card to the president of his bank with his account number, not his name. It was not *essential* to introduce numbers: the computers could have been programmed to deal with names and addresses. But using digits instead of alphabetic characters reduces data redundancy, and there is an ever-present pressure for such reduction as the new technology proliferates. The elimination of redundancy is one of the factors which can give the information age its cold and somewhat impersonal feel: numbers offer an efficient way of identifying individuals, but they do not resonate with our sense of identity.

The definition of a quantitative measure, entropy, marked a definitive step forward in clarifying some of the fuzziness that had surrounded discussion of information transfer. In the words of the great nineteenth-century physicist Lord Kelvin it is when science defines the measure of a phenomenon that it begins to understand and master it. However, this development also carried a sting in its tail.

For, though we now have an information measure that is

independent of the manner in which the information is represented, it is still dependent on the context in which the information is received. The fact that eight entropy bits of information are contained in the message that you are reading page 56 of this book no longer depends on whether you are communicating this in English, German or decimal digits – but there is a dependence on how many pages there are in the book, and also on any existing relevant knowledge the questioner might already have had. To be precise, it depends on the prior level of uncertainty as to which page you were reading – on his/her prior belief as to the information which is about to be received. If the questioner had any reason to narrow down his/her prior level of uncertainty – for example, s/he might have seen out of the corner of an eye that you were in the first half of the book – then the initial uncertainty level was not one part in 300 but one part in 150. The amount of information the questioner received when being told the correct answer was then fewer than eight bits, because s/he already possessed some of the total information.

Thus development of a rigorous definition and measure of information based on the knowledge previously held by its recipient brought back an unwelcome ghost which had haunted seventeenth- and eighteenth-century epistemology. The greater mathematical rigour which has been brought to bear on the evaluation of information has served to reaffirm rather than dispel the intrinsic subjectivity of the concept of knowledge. In a sense we have, after some 200 years of developing information theory, come full-circle back to the human mind as the basic seat and final arbiter of knowledge. The only context in which a person can interpret a new set of facts is the belief which he or she had before those facts became available – in the professional jargon, this is called the person's Bayesian Prior, after the eighteenth-century mathematician Thomas Bayes. If the previous belief is sufficiently strong – perhaps because of the accumulation of a sufficient number of contrary experiences, perhaps even from religious conviction – then it will be able, legitimately, to outweigh a great body of apparently objective incoming data.

Look at two successive editions of an encyclopaedia. What we accept as the body of human knowledge is changing all the time. As with an individual, the knowledge base of an entire society is subjective. With the passage of time, knowledge can never become a certainty, only an increasingly great probability, and never completely objective. Hume did not, in fact, lose the historical argument but won it; although his insistence that day may not follow night must be considered unusually obstinate, it is still a logically acceptable belief.

The data–information–knowledge hierarchy is illustrated on page 61. At the bottom is data, in the form of physical signals: light waves reflecting a scene being observed, or sound waves coming from a moving object. The natural data signals which reach human eyes and ears are analog. Their measure is frequency of oscillations. Since early history human civilizations have also developed digital symbols (alphabets and numerals) to represent words and quantities.

Next in the pecking order is information, or data interpreted as facts: the message coming through the medium. Overwhelmingly we distil information from data in the analog manner in which our brains naturally function, but since early history humans have also developed logical rules for manipulating digital symbols. The measure of information is entropy. Information reads a level of meaning into data – it interprets the physical signals of light or sound waves as something abstract, facts which the mind can work on later: 'this is a tree'.

Finally there is knowledge itself, information given meaning. Knowledge is the highest concept in the hierarchy and

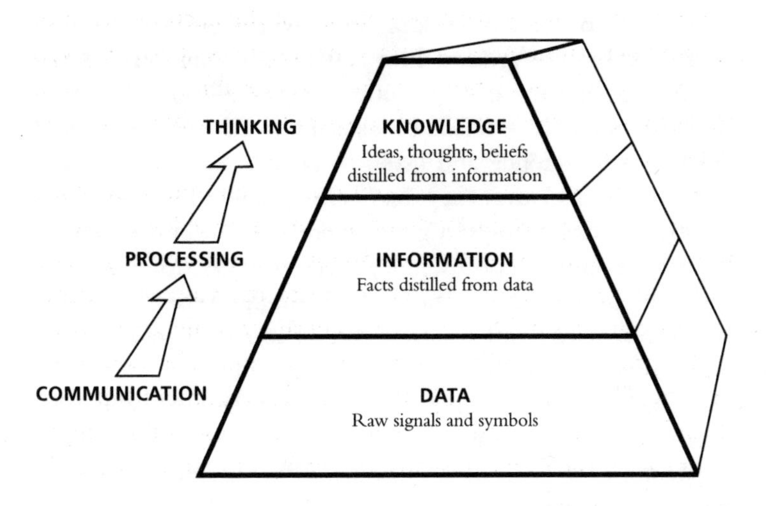

The hierarchy of data, information and knowledge, and the parallel hierarchy of communication, processing and information.

the most difficult to define; it includes the complete set of pre-existing beliefs of an intelligent entity – a person or society. Quantitatively its measure is the degree of certainty or uncertainty with which facts are known – or, rather, *thought to be known*, for it is always a subjective measure.

Paralleling these three levels are the three activities which humans engage in when working with data, information and knowledge. They engage in *communication*, the interchange of data with the outside world through the sense organs. They engage in *processing*, converting the physical data signals into recognizable images and facts. And they engage in *thinking*, as they mull over facts in the light of the accumulated knowledge in their minds, leading them to a new state of knowledge.

The information analysed by the hunter-gatherer standing amid the plains of Northern Europe 10,000 years ago was generated not by other people but, overwhelmingly, by the world of Nature around him. His descendants in the late twentieth century may be commercial lawyers, sitting at desks in skyscraper office blocks. They examine contracts, read correspondence and listen to clients' concerns. Nearly all of the information they analyse is generated by the minds of other human beings: it is received by them in the form of reports, letters, accounts, telephone conversations and the outcome of meetings, rather than through direct experience. It is man-made, and today much of it has been digitized. The growth of the base of human knowledge is decidedly speeding up. Scientific knowledge is doubling every decade; business information is arguably accumulating faster still.

Machines are helping us cope with this growth, at the same time as they help drive it forward. Until the nineteenth century the technology to help us see or hear at a distance, or to process moving images and sounds, did not exist. Then came society's two great information-technology projects: communications and computing. The goal of the first has been to overcome the distance barrier in human communication, to extend the natural range of our senses. That of the second has been to provide the capability to process information, to simulate some aspects of our mental powers. The realization of these two goals has been the story of the information age.

It began with communications.

3

Wiring the Planet

The biggest machine ever built is not an array of super-computers in the Pentagon, nor a nuclear energy facility or a particle accelerator, but something thousands of times larger than any of these. It is the global public tele-communications network. It has been growing for more than 100 years, since the first public telephone exchange was installed in New Haven, Connecticut, in 1878. At this moment more than a million engineers and technicians are working around the clock to extend and maintain it.

It is indeed a single machine, and each of us with a tele-phone has access to its controls. If you are by your phone right now you can select the number of a person in Indonesia or Peru and, by pressing buttons on your keypad, throw switches into action in exchanges around the world; a few seconds later that person's phone will ring.

The act of dialling is so routine that we forget the power of this machine we have the privilege of using. It is a remarkable feat of engineering and coordination. The feat is easily underestimated because so much of the machinery is hidden away or barely glimpsed. Each telephone wire leads to an exchange, a computer-controlled array of thousands of switching devices housed inconspicuously in a small building in a town or suburb. From there the signals pass from city to city along 'trunk' circuits capable of carrying hundreds or thousands of conversations simultaneously. In these circuits, and in the 'trunk' exchanges which link them to each other and to other circuits around the world, resides much sophisticated technology of which the public has only the vaguest awareness. Microwave towers dotted along intercity routes show where engineers have chosen radio communication as an alternative to laying long-distance, high-capacity cables. Satellite dishes point to orbiting retransmitters, which allow the signal to jump over even larger distances – across or between continents. In more and more cases it is making sense to take advantage of the enormous capacity of optical fibres, the latest development in transmission technology, which allows not thousands but millions of telephone conversations to be carried by a single cable.

A parallel achievement, with different historical roots but now integrated with this telephone network, has been global broadcasting. Many of the microwave circuits, satellite links and optical fibres carry a mix of telephony and television channels; cable television services, initially designed for delivering multi-channel entertainment, now offer many customers telephone services as well; cellular telephones and television broadcasts share the UHF frequency band. What began as a telephone network is now called the telecommunications system, and encompasses voice telephony, video and audio broadcasting, and, most recently, data transmission.

In 1988 Senator (later Vice-President) Albert Gore Jr.

coined the term 'Information Superhighway'. He conjured up an image of a high-capacity telecommunications network linking homes and companies throughout the nation and beyond. This vague but catchy term has become a metaphor for not only a technology – high-capacity digital networking – but also a new phase of the electronic age, in which businesses and citizens will have access to a previously unimagined quantity of information channels. The path of technology from the Morse telegraph and Bell's telephone to the Information Superhighway represents a long battle by scientists and engineers against the constraints imposed by the laws of physics, which place more obstacles than is commonly realized in the way of signals zapping across the country.

The industrial age developed machines which enhanced the muscle power of people and animals, but did not enhance our ability to hear or see. The speed of the oscillations which make up aural and visual data – from thousands to billions of cycles per second – make mechanical machinery quite unsuitable for simulating these functions. The development of information and communication technology had to await the emergence of a new science, electricity. This was a science of objects and forces which were free of the earthly constraints of weight and inertia – a science of the virtual world.

As creatures of the physical world we are slaves to Newton's laws. Each time we get out of a chair, we feel that to every action there is a reaction, that it takes force to accelerate a mass, and that the earth's gravity gives all objects that burdensome quality called weight. Likewise, the technologies developed during the Industrial Revolution were enslaved by these forces. The moving parts of any machine have mass and momentum, with all the limitations these qualities impose. Every driver knows that it takes more power to run a vehicle uphill than downhill, and that because of inertial forces a sharp bend has to be taken more slowly than a gradual one.

The electrons and photons in an electric wire or optical fibre are vanishingly tiny and weightless. There is no question of them slowing down to get up a hill or safely round a bend. The mass of an electron is one eighteen-hundredth of that of the lightest atom (hydrogen), which in turn has only a million-millionth of the mass of the smallest speck of dust which the eye can see. Photons move at the speed of light, some third of a million kilometres per second. On any scale of human perception, the transmission of a piece of information is effortless and instantaneous.

The twin phenomena of electricity and magnetism had been observed in Nature since ancient times, the former from mild electrostatic shocks and the attraction observed between pieces of amber when rubbed, the latter through the behaviour of naturally magnetized iron ores. The philosopher Thales studied such phenomena in Ionia around 600 BC and, with remarkable foresight, suggested a connection between them. The word *electric* was coined in 1570 by the English scientist William Gilbert, one of the pioneers of the study of electromagnetics. During the following 200 years very slow progress was made in the harnessing of this 'new' force, culminating in the announcement by the Italian scientist Alessandro Volta in 1800 that he had produced the first workable source of steady electric current – a battery known as a Voltaic pile. The development of the battery allowed rapid progress in the investigation and use of this new form of harnessable energy. In 1820 it was discovered – accidentally – that a current deflects a nearby magnetic needle, a property which not many years later allowed Samuel Morse to use an electromagnet to produce a paper record of the dots and dashes being sent over a length of telegraph wire.

The term 'electronics' is now almost synonymous with 'information technology'. Computing systems, as used in generations of traditional cash registers, were mechanical before they were electronic; the movement of fluids in microscopic pipes has also been exploited for computational purposes. However, these technologies did not become dominant. Electronic means of information manipulation became the technique of choice because it has two favourable characteristics which mechanical and other systems cannot match: ease of permutation and ease of propagation. Any medium which is to be used to encode and transfer information must be capable of producing permutations of its state (voltage level, mechanical position, or whatever) with great speed and efficiency if it is to cope with the very large quantity of data contained in human and natural information sources. No mechanical or fluidic system can offer anything approaching the millions of permutations per second achievable by electronic encoding, and the mere sub-billionths of watts of energy per second required by it. As for propagation characteristics, the electromagnetic means of sending energy is unmatched for speed and flexibility.

'What hath God wrought?'

This was the ceremonial inaugural message on the world's first telegraph link, opened between Baltimore and Washington, DC, on 14 May 1844. There was no doubt in the minds of Samuel Morse and the observers present concerning the stupendous significance of a technology which could convey information over an unlimited distance.

The Morse telegraph was the first effective electrical communication system. Like modern-day computers, it encoded each letter of the alphabet and each decimal number as a combination of two symbols. In Morse's code the two symbols were the dot, a short burst of current, and the dash, a long one. For generations of mariners *dot dot dot, dash dash*

dash, dot dot dot meant 'SOS'. The pulses of current, representing the dots and dashes of a message, travelled down the wire as surges of electrical and magnetic energy.

The next challenge was to develop a means of communicating not just the ons and offs of a finger on a telegraph key but the continually varying pressure waves of a human voice. An analog technology was required – one which could generate electrical waves in the telegraph wire *analogous* to the voice waves emitted by the conversants. Alexander Graham Bell patented just such a thing in 1876. His telephone mouthpiece detected sound waves in the atmosphere and, like the ear, converted them to electrical signals. Whereas the human ear's signals travel along the auditory nerve to the brain's primary cortex, Bell's mouthpiece sent the signals along copper wires to another device, a telephone earpiece, which reproduced the sound waves that had been detected by the mouthpiece. For the first time the interchange of aural data between humans was mediated electronically, overcoming the natural barrier of distance. The initial telephone message – Bell's instruction to his assistant, 'Come here, Mister Watson, I need you' – travelled only the length of the laboratory, but within a few years the messages would travel between cities and countries. The age of telephony had begun.

Morse and Bell were inventive geniuses but not men of science (indeed, Morse was by occupation a house-painter), and did not seek to supply a theoretical basis for the methods they used to communicate along electric wires: all both men were really interested in was that their devices *worked*; they were not much concerned with the *why*. And their methods, pioneering as they were, are not the methods used in the global telecommunications network of today. The original telegraph signals operated as would a child's electrical set, making the buzzer buzz at one end of the room by opening or closing a switch at the other. Such a signal may get around a house or down the street, but it will not get far around the globe before disappearing in a welter of electrical distortions and interference.

In today's networks the electrical energy propagates as oscillations – waves of energy. These are tuned to the resonant frequencies of the medium through which they travel: a coaxial cable, an optical fibre, a microwave radio link, or whatever. Nevertheless, designers of modern computing and telecommunications systems, with clock speeds of millions of cycles per second and upwards, face the same obstacles in terms of the behaviour of electrical energy as it flows along wires or through the airwaves as were confronted by their nineteenth-century predecessors.

Further progress was to require a greater understanding of the science of electricity and the propagation of electrical energy. This understanding was to move forward rapidly in the second half of the nineteenth century. Credit for its definitive development goes to the great physicist and mathematician James Clerk Maxwell, who in 1878 published four compact equations which are the laws of electromagnetism. Electronics is not a slave to Newton's laws, but it is to these.

In *The Emperor's New Mind* (1989) Oxford Professor of Mathematics Roger Penrose divides the theories of science through the ages into three categories: Superb, Useful and Tentative. To qualify as Superb, the range and accuracy with which a theory applies to the world should be, in his word, 'phenomenal'. Only two theories from the ancient world and four from modern times satisfy his tests. The first ancient one is Euclidean geometry (a straight line is the shortest distance between two points, parallel lines never meet, and so forth); only in the twentieth century has Einstein's General Relativity given us a more precise description of the geometry of the space in which we live, and even so the

Universe's departure from Euclidean flatness due to the curvature imposed on space by gravity is tiny – less than the diameter of an atom over a distance of a metre. The second Superb theory from ancient times is the theory of statics developed by Archimedes and his contemporaries, of leverage ('Give me a fulcrum and I will move the world'), buoyancy and other forces between stationary objects.

The four modern Superb theories are Newtonian mechanics, Maxwell's theory of electromagnetic radiation, Einstein's theory of General Relativity, and the 'strangely beautiful and revolutionary' quantum physics. With one exception, these great advances in basic science are deeply etched in the minds of the educated population. Newton's laws have created a sense of order and causality in the physical world. Einstein's theory represents the new science of the macro-world, of galactic distances and cosmic energy levels, encapsulated in the public mind in his equation $E = mc^2$. Quantum Theory addresses the even more mysterious micro-world of the internal structure of the atom, where determinism breaks down and is replaced by a principle of uncertainty.

Of Maxwell's laws we hear less, but without a grasp of them it is impossible to get a feel for the infrastructure of the information age. These laws show that, in addition to the familiar world of tangible objects, there is an intangible world of force fields which can propagate as waves through empty space. Depending where they are in the electromagnetic spectrum, these waves can be radio, radar, heat, light, X-rays or gamma rays. Maxwell's equations clarified the lack of certainty surrounding the behaviour of electric and magnetic fields which had made the development of early telegraph and telephone systems somewhat unscientific.

Like many great theories, the new picture of the natural world which emerged from the study of electricity and magnetism in the late nineteenth century had a very simple foundation. This is that matter has, in addition to the familiar property of mass, a further property called *electric charge*. Both are fundamental or axiomatic: there is no theory or explanation of them; they are simply basic properties we must accept as *there*.

Electric charge, this newly discovered basic property of matter, was found to manifest itself in two ways: as flows of electric current, and as radiation of electromagnetic waves. Electric currents arise because, in a conductor like a metal as opposed to an insulator like rubber or plastic, some of the charged particles (electrons) within the atomic structure of the material are not tied to individual atoms but are free to wander. They move along a conducting wire if a voltage (a difference in electrical energy level between one end and the other) is applied to it. So long as the current flows steadily and in a single direction, that is essentially all there is to the transmission of electrical energy.

If the electrical current varies in any way, as in the alternating current (AC) supplies of commercial and domestic power, the pattern of flow of electrical energy becomes more complicated. It generates waves of energy. An alternating current in a wire sets up around it an area of force known as an electromagnetic field. A field is an area of force in the sense that, if another electrical charge (or a magnet, which is why the term 'electromagnetic' was adopted) moves through it, then a force will act upon the electrical charge (or the magnet). This force can be put to use in electric motors, loudspeakers, bells and the like. At the relatively low alternating-current frequencies of electrical power systems (50–60 cycles per second), these electromagnetic effects are very weak unless bolstered by the presence of a material with suitable magnetic properties, usually iron. The iron cores around which electrical wire is wrapped in electric motors, loudspeaker drivers and transformers magnify greatly the interaction between nearby electrical conductors. As a child playing with magnets knows, the magnetic field emanating from these objects (which is a static version of the changing magnetic field of an electric motor or transformer) does not carry far through air or other non-magnetic materials.

Varying fields of electromagnetic energy are the means by which information is transmitted in a communications network. Over the following 100 years a great variety of communications technologies was developed, always of two types: those in which the signal travels freely through the air, in the form of radio waves, and those in which it is guided along wires, cables or optical fibres. If the waves travel through the air, the engineering task is to make them propagate as far as possible; if they are guided along wires, the task is to make them propagate away from the cable as little as possible.

A large office building looking over the city and lake of Geneva houses a branch of the United Nations called the ITU – the International Telecommunication Union. Its function, to coordinate the transmission of telephone, television and other signals across frontiers, is largely technical. The hundreds of engineers it employs carry out their tasks unencumbered by the political controversies which bedevil other United Nations bodies.

With one exception. At intervals of several years the ITU holds a conference to carve up the electromagnetic spectrum. The WARC – World Administrative Radio Conference – is attended by officials of every country in the United Nations. The resource over which they squabble is the ability to use the airwaves for the technologies of their choosing. The conference decides which frequencies can be used for what purpose – such as television, cellular telephony or microwave links – and, when the radio waves are of sufficient wavelength to cross national frontiers, which frequencies can be used by which countries. The decisions have the force of international law, and their economic impact is big; in the United States alone, auctions of frequencies for broadcasting or cellular radio use have yielded billions of dollars of federal revenues.

Telecommunications had first taken to the airwaves in 1898 with the public demonstration by Guglielmo Marconi of a wireless telegraph; he provided the *Dublin Express* with live coverage of the progress of yachts in the Kingston Regatta, which he observed from a boat as they raced in the Irish Sea. Earlier Marconi had sought assistance from the Italian Minister of Posts and Telecommunications for the development of his invention, but elicited no interest. The chief engineer of the General Post Office in London, however, recognized its potential, and the young Marconi moved to Britain, in due course to create the telecommunications equipment company which still bears his name. For the next twenty years radio as a means of telegraphic communication spread extremely rapidly around the world; Marconi made the first transatlantic transmission on 12 December 1901, only three years after his sports broadcast to the *Dublin Express*. Telegraphy played a critical role in World War One, during which military commanders had the novel ability to receive news instantaneously from, and give orders instantaneously to, battalions in the field.

Early commercial radio communications operated low down in the long-wave band, at frequencies of under 100,000 cycles per second. (The length of a wave is the distance it travels before repeating itself, which gets shorter as the frequency of oscillation gets faster.) That was as fast as contemporary electronic circuitry could generate the oscillating currents which launched the radio waves on their way. These slowly oscillating waves also had the merit of travelling (propagating) well. Later the medium- and short-wave bands were opened up.

What has stopped the allocation of this valuable international resource degenerating into open dispute, or worse, has been the march of technology. Such is the demand for global information-carrying capacity that, with the passage of time, frequency levels are continually being pushed upwards. In the middle decades of this century the frequencies used went even higher. First came Very High Frequency (VHF) signals, used for higher-quality radio and for early television broadcasting. Then Ultra High Frequency (UHF) technology was developed for superior television broadcasts and for mobile radio. Cellular telephony was launched at 450MHz and then moved up to 900MHz. When the speed of oscillation reached

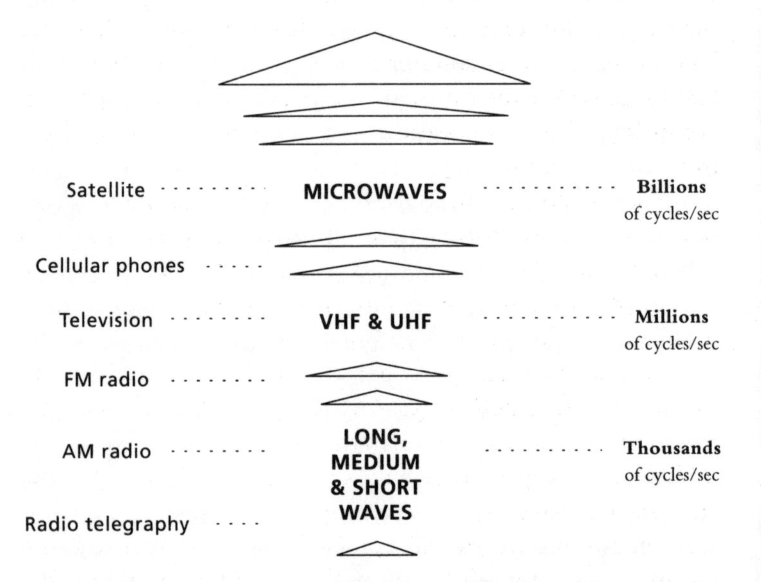

The electromagnetic spectrum: with the passing of the years, science opens up higher and higher reaches of the airwaves to cope with the explosion of applications we demand.

1,000 million cycles per second (1GHz) the adjectives began to run out. The term 'Super High Frequency' was adopted briefly, but gave way to the less cumbersome word 'microwaves' – a reference to the fact that, at this speed of oscillation, the waves are very short, only a few centimetres long.

Huge increases in information-carrying capacity accompany each extra step up this ladder. Roughly speaking, each new band – long, medium, short, VHF, UHF, and so forth – adds ten times more information-carrying capacity than all the previous bands put together. The progression is exponential: add two new bands and the total capacity is increased one hundredfold; add three new bands and it is increased by one thousandfold. The constraint on opening up the spectrum yet further is not in the laws of physics: there is nothing about Maxwell's laws which prevents electromagnetic waves oscillating millions of times faster than any communications system has yet caused them to. The problem we encounter is a practical one: as the oscillations get faster they present greater difficulties in terms of controlling their propagation – in our ability to guide them from transmitter to receiver.

Relying as much on intuition as on science, the working out of how an electromagnetic signal propagates through an urban or natural environment is an arcane art, occupying the minds of thousands of engineers designing broadcasting, mobile and other communications systems. Roughly speaking, waves can manage to get around obstacles which are smaller than themselves. An AM radio signal in the medium or long wave, whose wavelength is hundreds of metres, will diffuse around the largest city building. A UHF television antenna has to be more carefully placed; at a few metres wavelength, the waves will get around trees or a garden shed but may have more difficulty with a large building. Users of cellular telephones, which also operate in the UHF band, are familiar with this phenomenon: one side of the street may give a clearer sound than the other. The highest-frequency signals which domestic users are likely to encounter, microwave signals received from a satellite, are only centimetres long; they require an almost clear path (without obstacles such as buildings) through the airwaves. The propagation characteristics get steadily less favourable as the frequencies increase: in all of these cases the earlier, longer-wave systems required fewer transmitters to give national coverage, and penetration around natural obstacles or into buildings was better.

Nevertheless, the added capacity of the higher frequencies is worth paying for. Microwaves are particularly suitable as communications carriers because of the very large range of frequency available. The characteristic dish of a microwave antenna can be seen on innumerable commercial building roofs and on towers dotting the countryside. Microwaves provide the main backbone for long-distance communication; they have become especially key since the development of the communications satellite.

Conceptually, a communications satellite is simply a microwave relay tower placed in space. The idea that radio waves could be beamed up to such a craft and sent down again to other spots on the earth's surface was conceived by the science-fiction writer Arthur C. Clarke in 1945. However, it was not until the Russian–US space race had brought to perfection the skills of launching spacecraft into earth orbit that this technique became commercially viable. The first commercial communications satellite, *Early Bird*, entered service over the Atlantic in 1965. It enabled for the first time live television coverage of events in Europe and North America to be shown to viewers in both continents, bringing awareness of this new communications technology dramatically into homes on both sides of the Atlantic.

In practice, an orbiting spacecraft is in a different league of complexity to a terrestrial microwave tower. It needs to position itself with great accuracy in its orbital position using small rocket motors, and to readjust that position periodically during its design life of approximately ten years. After reaching its slot in space it must unfurl the large panels containing the solar cells which will provide its electric power; these are the wing-like structures which give the satellite its characteristic appearance. The actual transmitting and receiving devices, known as transponders or simply as channels (each might accommodate a television channel or hundreds of telephone circuits), are pointed at areas of the earth's surface which may be the size of a continent. Such an area is known as the footprint of the transponder: it is where the signal is strong enough to be picked up by a receiving dish.

All this technology – the bus (which is the name given to the spacecraft itself with its rocket motors and fuel supplies), the twenty-four or so transponders and the wing-like solar-cell arrays – must weigh less than a tonne and be able to withstand the stress of a full-scale space launch. The cost of building each satellite and putting it into space exceeds 100 million dollars, every cent of which is lost if the launch operation fails – as has happened many times.

Despite these complexities and risks the communications-satellite business has become a major industry; it is the most successful commercial product of man's space endeavour. The development of these satellites was a dramatic step forward in the communications world. A satellite in the sky creates at a stroke the equivalent of an entire network covering an area the size of one, two or even three continents (it cannot point transponders at all the continents from a single geostationary orbital position). A receiving (or transmitting) dish can be placed anywhere in the footprint of the satellite and be connected to the world's communications systems. No prior technology 'shrank the world' as did this one. It also gave telecommunications and broadcasting technologies a new image. In the 1970s, when telephone and television corporations joined the space age, every company brochure contained photographs of satellites in space or parks of dishes on the ground. The satellite became the icon of a new era in communications: instantaneous, global, ubiquitous.

Those early satellite dishes of the 1970s were huge objects: referred to as earth stations (the satellites were the corresponding space stations), they cost a quarter of a million dollars per dish. The technology could be used only for very high-value applications, principally to provide inter-continental carriage of telephone and video circuits. Subsequently dishes became smaller and cheaper, due to advances in amplifier design which allowed weaker signals to be satisfactorily detected, and by the late 1970s they were within the price range of cable-television operators, who were therefore able to provide their subscribers with a greatly expanded range of services. The major expansion of the cable industry in the United States dates from this time. Ironically, further reductions in dish size and cost then produced the greatest threat to the cable industry: viewers began to buy their own dishes, bypassing the cable network. By the early 1980s satellite receivers were about two metres across and could be mounted in a back yard; by the late 1980s they were sixty centimetres across and could be fixed to a window frame.

Telephone applications benefited from the same technological advances. Companies routinely use dishes mounted on their corporate headquarters to connect straight to the world's long-distance communications paths, bypassing the local telephone company infrastructure. Even mobile telephony makes use of this technology. Commercially available portable telephones now enable a user to be in two-way telephone contact with most other points on the earth's surface.

★

There was one more step to go in the search for higher-capacity communications.

A nineteenth-century family-run glassworks in the foothills of the Appalachian Mountains may seem an unlikely setting for one of the key breakthroughs of the information age. But the Corning Company has a long pedigree of high-technology patents – it made the first Edison lightbulbs, pioneered the mass production of tubes for TV sets, and is the world's largest producer of the glass substrates for the flat screens of personal computers. The innovation which has now put Corning at the very heart of the telecommunications revolution was optical fibre. The company owns the key patent for the production of the very pure glass fibres required for the long-distance transmission of light.

The first demonstration that a fine fibre of glass could be used as a conduit to guide light waves was made in 1955 by the Indian-born scientist Narinder Kapany. Strictly speaking, the use of light instead of other electrical oscillations for encoding and transmitting information represents merely a movement further up the electromagnetic spectrum: light is just another kind of radio wave, differing from the rays emanating from a microwave tower or cellular telephone only in its very much higher frequency. In practice, however, the propagation characteristics of light in a fibre are radically different from those of electricity in wires or radio waves in the atmosphere.

Optical fibre is the latest in a series of technical improvements in the conduits used to wire up homes and offices. The original and cheapest form of wiring for telecommunications purposes is a pair of copper leads. This is still the wiring most commonly used throughout the world to provide telephone services. The frequencies of electrical oscillations generated by a telephone handset are exactly those generated by the human voice, up to several thousand cycles per second. At these low (by electronic standards) frequencies most of the

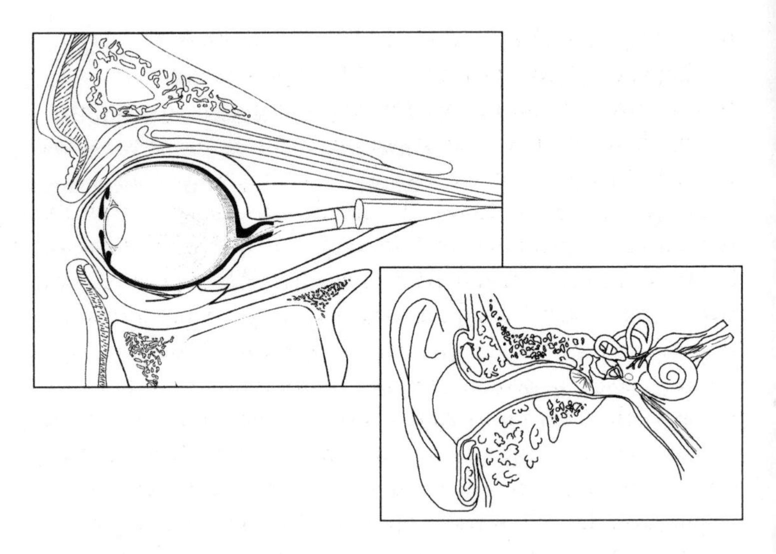

The human eye and ear are exquisitely refined communication sensors; they are the standards to which a telecommunications network must aspire if it is truly to 'kill distance'.

energy stays in the wire; only a little leaks out beyond the insulation, occasionally to produce the annoying phenomenon of cross-talk between two telephone lines which happen to lie near each other in an under-street duct.

The human eye proved more difficult than the ear to simulate electronically; that achievement had to wait until 1926, when the Scotsman John Logie Baird invented television. Baird focused the image to be observed onto an array of sensors which, like the rods and cones in a human eye, detected light intensity and converted this into electrical signals. Baird's pictures were crude by today's standards: flickering black-and-white images on a small round screen. But the foundations had been laid, and by the 1950s the use of leisure time throughout the Western world was being transformed by the ability to watch live coverage of events and entertainments on affordable television sets. Now full sound and vision – those two overwhelmingly dominant human sensory channels – could be transmitted.

The pairs of copper leads used to supply telephone service were, however, unsuitable for transmitting these new television channels. The quantity of information in a video signal exceeds that in a voice signal by an enormous multiple, typically a factor of 1,000: the bandwidth picked up by the human ear is measured in kilohertz, but that of a television broadcast signal is measured in megahertz. This problem was overcome (when 'wired', as opposed to 'over the airwaves', transmission of television pictures was called for) by using coaxial cables, in which the pairs of conductors are not next to each other, as in normal telephone or electrical wiring, but one inside the other. The signal travels in the central conductor and is sheathed, or screened, by the outer one. It required a relatively simple application of Maxwell's equations to show that such a screening arrangement vastly reduced the tendency of the energy to dissipate as it travelled along the cable; the electromagnetic energy was contained within the cable by the outer screen. The passing of signals along a coaxial cable has some of the characteristics of electrical conduction (electrons flowing along wires) and some of the characteristics of a radio broadcast (energy waves travelling through the area between the inner and outer conductor).

With the increasing popularity of television in the middle of the twentieth century, cables began to be laid in urban areas for the purpose of supplying homes with a clear picture from a single high-quality antenna. These were the first cable television systems. Coaxial cables, physically much thicker than twisted pairs, allowed transmission of vastly more information than a telephone wire, and, most importantly, the transmission of video. Just as the optic nerve leading from the human eye is much thicker than the auditory nerve from the ear, so the man-made wiring for television had to be of much higher capacity than the wiring for telephony.

The optical fibre represents a great step forward from the use of telephone wires or television cables. Having gone through several technological generations of improvement since its first commercial use in the 1970s, it is now close to perfection as an information conduit. Energy loss through the boundaries of the fibre is so low (it cannot be zero, for electromagnetic radiation can never be totally contained) that the signal can travel hundreds of kilometres without amplification. Cross-talk and other aspects of interference with neighbouring conductors is essentially nonexistent: there are none of the problems electrical systems have with unwanted radiation of energy. The light injected into one end of the fibre is pure (single-frequency), and that is how it emerges at the far end.

Above all, fibre has phenomenal information-carrying capacity. The frequency of light is so high that the capacity of a fibre is difficult to imagine. A million telephone conversations can travel simultaneously along a single fibre of diameter one-tenth that of a human hair. Such fibres are now being introduced into numerous homes and offices throughout the world.

The development of optical-fibre systems has overcome the constraints of electrical encoding which troubled telecommunications engineers for the previous century and a half. While other communications breakthroughs in recent decades, such as satellite technology and cellular communications, have had outstanding consequences, optical transmission occupies a different level in the scale of innovations because it is the first information technology to go beyond electronics. If electronic transmission is like a bicycle, optical communications is like a space rocket: bicycle and rocket are in theory just different transport devices, but in practice they are capable of vastly different levels of performance. The introduction of optical fibres has lifted information transmission to a level of performance which quite simply takes the matter of capacity limits out of the range of concern for the designer of a network.

A century and a half after telegraph systems started to criss-cross the American continent, there are still only three types of wiring in the world's telecommunications systems: twisted pairs, coaxial cables and optical fibres. They have fundamentally different properties as carriers of waves of electromagnetic radiation. The capacity of the twisted pairs is measured in kilohertz, the coaxial cable in megahertz, and the optical fibres in gigahertz.

If telephone companies had laid coaxial cables instead of twisted pairs for their subscribers during the last several decades the story of the transition to the world of multi-media entertainment and high-speed Internet access would now look very different. But that was, obviously, not the priority at the time. Consequently the typical home in North America, Europe or elsewhere in the developed world now has two electrical connections to the communications infrastructure: a telephone wire, through which access can be gained to practically any other phone in the world but which is restricted in its information capacity; and a cable-television connection which has enough bandwidth for some 100 full video channels but carries signals one way only, downstream to the home – and also has no capability to switch from one communications source to another.

Finally there is fibre, the uncontested solution to almost any telecommunications task, permitting operations at the speed and oscillating frequency of light. The only obstacle is cost: these days fibre is not greatly more expensive than a copper pair, but multiply the difference by millions (for the millions of customers a telephone company has) and the investment is difficult to justify if people can make do with the old wires. Even if the cost becomes the same, replacing the old wires will require an investment of hundreds of billions of dollars.

Digital communication produced a new set of problems for telecommunications engineers. A bit stream – a series of clean transitions between 0s and 1s – is a mathematical construct, a product of the human imagination, which cannot exist in the physical world. The laws of Nature do not allow it. Nature is analog; whether an animal is calling or an electromagnetic wave is moving through the atmosphere, Nature's communications consist of oscillations, waves, cycles.

A digital bit stream can be approximated but cannot be exactly replicated. At low frequencies, the approximations are adequate. When you walk into a room and switch on the light, the current in the wires does not build up instantaneously but over time; the change is complete within a fraction of a second, an insignificant delay if you just want to see what is in the room: as far as you're concerned, the light has come on instantly. Likewise, over the short distances and slow bit rates of nineteenth-century telegraph links, these problems of delay could be ignored; the electrical energy was largely confined to the wires and the loss through radiation. A very primitive link like Samuel Morse's 1844 telegraph worked by just switching on and off the current in the wire; the small fraction of a second which it took for each bit to build up and decay could be ignored.

Today's networks have to do better than that. The clock speed – the rate at which a digital system processes and sends out data – is nowadays so high that the beginnings and ends of each bit would merge into each other and, not far down the wire, would become unintelligible. Furthermore, any changing current of electricity generates waves of electromagnetic energy which are not so confined; they can radiate out to the surrounding environment. At the low frequency of changing current at which traditional power systems operate – 60 cycles per second in North America and 50 cycles per second in most of the rest of the world – the extent of energy leakage through electromagnetic radiation is very small. We hear some humming in high-voltage power lines and transformers as a by-product of the interplay of electric and magnetic fields, but the overwhelming proportion of the energy transfer is via the movement of charged electrons within the confines of the conductors. However, at the high clock speeds of modern computing and telecommunications devices – millions of cycles per second and upwards – the transmission of bits along wires is perhaps better visualized in terms of travelling waves of electromagnetic energy than as the simple movement of a current of electrons.

Two enemies thwart the transmission of energy waves down a wire: resistance and reactance. Resistance is the failure of metal to act as a perfect conductor; when charged electrons move along the metal there is dissipation of energy in the form of heat, the body of the wire warming up slightly. Resistance is the main source of loss in high-power transition systems – in national and local grids. In information-technology equipment its effects are less pronounced, but nevertheless must be taken into account.

Reactance causes bits to become distorted in shape as well as reduced in amplitude. This is the more fundamental problem because it ultimately makes the bit stream unrecognizable, and it cannot be corrected by periodically reamplifying the signal. Reactance is the unwillingness of Nature to accept too rapid a change in the current flowing in a wire; it will certainly not accept a clean transition from on to off, but releases energy waves into the surroundings as if in protest. Yanking a plug of a live socket will produce a spark. Just switching a room light on can generate an audible interfering click in a nearby radio.

The problem is overcome by converting the bits into tones – analog tones – so that they can travel freely as oscillations, waves, cycles. A communications link, be it cable, satellite or optical fibre, is an instrument which has to be played. It has frequencies to which it is tuned, to which it will resonate so that it can carry waves to their destination. Other frequencies will not be carried. Digital bits, by contrast, are not waves at all, so they are not in tune with any medium.

The most familiar device for achieving this conversion is the modem, a MODulator–DEModulator. This in effect whistles down the line, modulating the whistling tone a little lower for a zero, a little higher for a one – or, in the more sophisticated modems around today, whistling quite a few different tones depending on the combination of bits being sent. Sending data via a modem is like playing the keys of a piano. Each key has different meaning, and a listener can hear the sounds and interpret their meaning. Sending unmodulated bits down a wire is like thumping the case of the piano. The dull thuds could perhaps transmit a rudimentary Morse code message, but they would not be heard far and would not do justice to the instrument as a means of communication.

The digitization project of the telecommunications industry has as its goal the conversion of all links in the chain between any two users in the world. The digits travelling these highways would represent computer data or, often, voice or video in digitized format.

A modem is the crudest in the hierarchy of technologies for digitizing a line. The frequencies it produces are way down in the audible range – hence the squeaky pipping noises made by computers and fax machines as they make their connections over the telephone network. The equipment used in the major links of national and international networks make their oscillations millions or billions of times

The Information Superhighway is a metaphor for the medley of cable, microwave, satellite, radio and optical fibre links that will carry the digitized telecommunications traffic of tomorrow.

per second. These waves are tuned to the natural resonating frequencies of the physical medium, the frequencies at which each medium can carry them: for a coaxial cable anything up to several hundred million, for a microwave or satellite link several billion, for an optical fibre many trillions of cycles a second.

It is to this concept of universal digital communication that Al Gore referred when coining the term 'Information Superhighway'. Integrating through digitization the individual developments which preceded it – of telephony, radio, television, satellites, optical fibres – the Information Superhighway brings together these individual technologies into a network which represents a culmination of the century-and-a-half-long communications project.

It will not actually be a uniform physical network. The surface of the earth is not flat: it has mountains and rivers, continents and oceans. The physical links, be they copper cable, fibre or microwave, are matched to the physical topography which they must traverse. Some technologists dream of an all-fibre-optic world network, with its practically infinite capacity being available uniformly to all users, but this is not financially realistic. It would not even be optimal, because it would fail to adapt to the natural advantages of each medium: radio connections, while having lower bit-rate capacity, provide useful extra mobility, while the ability of one satellite signal to cover a whole continent makes that technology well suited for broadcasting.

When the project is complete, every link in every network will carry bits – uniformly and free of errors. In not many years from now, fully digital communication will be the rule. We have to qualify that statement slightly because at their core the networks will not be truly digital; a digit is only a mental construct and nothing (that is, no thing) can actually be digital. But they will be as digital as the laws of physics allow, and the analog oscillations will be at frequencies far above the audible range. The whistling and humming of modems will stop and, for practical purposes, only bits will travel.

We are not far from the completion of this task. Between the 1960s and the 1990s most inter-exchange, intercity and international links were converted to digital operation. This investment cost billions of dollars worldwide. The final obstacle is, curiously enough, the so-called 'last mile' between a local telephone exchange and the user's telephone. This is proving a tough nut to crack. The 'last mile' almost invariably uses a twisted pair of copper wires – the worst kind of medium for carrying rapidly changing data. The length of each such connection may be literally only a mile, but there is a separate one for every telephone line in the world – some 700 million at the last count. Nevertheless, conversion is coming. Large office buildings already have digital access to the telecommunications infrastructure, as do many houses and every owner of a digital cellular phone.

The fact that there is now a finish in sight for the communications project is of great significance in understanding the information-technology revolution. The endpoint is a generally accessible, broadband integrated digital service. (*Broadband* means having a capacity which can accommodate any commercially available input or output devices which the user may wish to connect. *Integrated digital* means integrating into a uniform digital bit stream the previously separate technologies of telephony, broadcasting and data communications.) This will be the Information Superhighway. Every separate feature of it – general accessibility, broad bandwidth, integration, digitization – already exists. They will soon be brought together.

When the Superhighway project is complete we will all discard our modems, because they will be history: the links themselves will be digital. To be more precise, since what the links actually carry are always analog energy waves, they will be digitized by the operators – the telephone companies, cable-television operators or other network providers.

The rate of announcement of new telecommunications solutions is on the surface bewildering. Cable-television companies, blessed with an infrastructure capable of carrying broadband signals, are beginning to offer digital access at many megabits per second. The telephone industry is fighting back with a technology which carries data over normal telephone lines at speeds of up to or in some cases more than a megabit per second. Meanwhile Teledesic, a consortium comprising the aerospace company Boeing, the telecommunications carrier McCaw and the software producer Microsoft is set to bypass the earth-bound networks altogether by launching hundreds of low-orbiting satellites to provide a multi-megabit service to any user in any location around the world.

However, when the details are stripped away a simple pattern comes through. The underlying carriage is either:

- one of the three competing technologies for cabled connection – copper pair for multi-kilohertz, coaxial cable for multi-megahertz, or fibre optic for gigahertz capacity, or
- over the air.

For the latter, the frequencies now being exploited commercially are mainly those in the upper part of what has traditionally been thought of as conventional broadcasting territory (UHF and VHF), now shared with cellular telephony, or, higher, in the enormously high-capacity microwave territory.

While digitization allows for great flexibility in the ways in which a communications link can be used, there remains the absolute requirement that a physical (analog) path must be used as the carrier which will take the bits from A to B. This fact is determined by absolute scientific constraints. The provider of any proposed telecommunications infrastructure must either obtain use of that valuable resource called the electromagnetic spectrum or physically lay the requisite cables. The economic value this places on the spectrum option is increasingly being realized by governments, who several years ago began the practice of selling it. The buyers – telecommunications companies – know that, despite the transition to digital coding of information, the 'real estate' of analog bandwidth will always be the key underlying resource. It will continue to be fought over through auctions at the local level and, at the global level, through diplomacy at the ITU.

Technologically, society has now cracked the communications problem. High-quality sounds and images can be transmitted instantaneously along wires or through the air. The data inputs to the human body, the bottom level of the data–information–knowledge hierarchy, can be simulated electronically to as high a level of resolution and sensitivity as the sense organ itself can muster, can be encoded digitally, and can be transmitted essentially error-free to any destination on the surface of the earth. The only sensory modes which we transmit in quantity are vision and sound, as these high-bandwidth channels dominate person-to-person communication; but our networks could equally transmit digitally encoded smells or tastes or pressure readings if the user wished to connect the requisite apparatus at each end of the communication link. In that respect the achievements of the early pioneers of the communications age – Samuel Morse's harnessing of electricity to send a message by wire, Marconi's harnessing of radio to do so through the ether – have been brought to completion.

With his four equations describing how electromagnetic fields propagate, Maxwell provided a rare example of a theory which in philosophical parlance is *closed*, the equations each complementing each other and leaving no factor unexplained – which is why Penrose ranked Maxwell's equations alongside the achievements of Euclid, Newton and Einstein in the ranking of science's great theories. So long as the mode is electromagnetic, Maxwell's laws have symbolically closed the debate about what is and what is not possible in communication – what signals will and will not fit into a copper wire or a sliver of the electromagnetic spectrum.

What is left to achieve is for the technology to have wider coverage at lower cost. The networks created during the course of more than a century to attain this have bequeathed us a distinctive technical inheritance which we must, for better or worse, accept as the basis from which to build communications for the future. As technology designers look to the twenty-first century they are acutely aware that the majority of computers are still communicating with each other (if at all) along circuits which date from a much earlier era. The global telephone system, still the backbone used for most computer communications, was designed to carry the analog sound of the human voice, not high-speed streams of digital data. Most homes still have as their only link to this global network a pair of thin copper wires laid decades ago. Here the constraints of bandwidth present a difficult barrier. With the advance of technology and the construction of higher-capacity links increasing the bandwidth available, that barrier will be surmounted.

4

The Chip, Master Logician

The theme of the first information revolution was communications – the telephone invented by Bell in 1875, the radio by Marconi at the turn of this century, television by Baird in 1926. Developments in related technologies helped to create an entertainment industry based on electronics: the recording of sound on records and tapes, the amplification of voice and musical instruments, and the development of high-fidelity reproduction equipment for domestic use.

These earlier, pre-computer-age technologies were of huge economic and social consequence. In Marshall McLuhan's famous phrase, they brought about the global village. The telephone is still the essential tool of every business: it is the item business managers would give up last if every piece of technology, certainly including the computer and probably even the electric lightbulb, were to be stripped one by one from their offices. As for watching television, statistics confirm that it now has no challenger for the number one position as mankind's leisure pursuit. Many consider that it is equally dominant as a former of attitudes and preferences.

All these technologies were analog. As the name implies, the electrical currents are an imitation, a copy, an analogy, of the sounds and images being transmitted. Whistle a middle C and you produce pressure waves in the air at 262 cycles per second. Whistle it into an analog telephone and the electrical currents in the wires to the exchange will actually oscillate at that same 262 cycles per second. At any point along the wires an amateur sleuth with headphones could touch into the circuit and would hear that whistle. It is there, raw and unprocessed, all the way along the transmission path. Similarly in a radio broadcast the electromagnetic waves rise and fall in amplitude or frequency in precise synchrony with the modulations of the presenter's speech. A child with a piece of wire, a coil and a 10p diode could pick out the programme; many budding radio technicians have done just that over the decades. Analog communications systems, even technically advanced ones such as cellular telephones, are notoriously easy to tap.

As impressive as these analog technologies are, there is something about them which has failed to capture the modern imagination. There is a feeling that the world of computers and satellites, of information networks and of virtual reality, is more awesome and powerful than the analog world which preceded it. The ingredient missing from analog systems is *processing*. With a digital system, unlike an analog one, you can actually work on the content or meaning of the information. The system can take in a word, study it and put out at the other end a different word, not just a faithful imitation or reproduction of the input word. Terms like 'meaning' and 'content' must be used carefully here; there are different levels of meaning, the higher of which a computer cannot reach in its interpretation of data. But the fact that digital systems can interpret content *at some level* – can examine it, process it and change it in accordance with the instructions of the user – is fundamental to their character and their significance. It is what makes them able to enhance the mental powers of their users.

With digital technology we create a world of coded information. This means the information is no longer in the raw form which Nature presents to our eyes or ears, but is encoded in accordance with rules of arithmetic and language. Quantities are represented by numerals, by decimal points and fractions. The numeral 1,000 means a lot of something and the numeral 0.001 means very little of it. A digital computer, appropriately programmed, will recognize that the larger number is a million times greater than the smaller one. Letters, too, are digital code. H, O, U, S and E, in that order, means something large and inanimate, and M, O, U, S and E means something small and animate. The computer could correctly classify the house as a building and the mouse as a rodent. Our ability to handle concepts through structured language (words with spelling and sentences with syntax) and to handle quantities through structured number systems (numerals which can be added, multiplied and even raised to powers) is one of the defining characteristics of the human species. Apes, dolphins and other higher animals can associate individual human words with meanings, but cannot interpret decimals. The digital system interacts with humans on the higher intelligence plane of encoded language and numerals, not on the more animal plane of natural sounds and images.

Computing – the development of machines to process information digitally – constituted the second of the great information-technology projects.

Alan Turing is known in the popular mind as the code-breaker *extraordinaire* who helped win World War Two by cracking the German Enigma cipher, and as the brilliant but psychologically tortured man who, persecuted for his then illegal homosexuality, committed suicide in 1954. In the history of ideas he will be remembered for the Turing machine.

As a young mathematician studying for a doctorate at Princeton University in the 1930s, he became interested in whether problems of logic could be solved by mechanical means. If a puzzle is capable of being solved by the application of human reason, what kind of machine could exist that would likewise be able to solve it? A Turing machine is (to simplify slightly) a read–write head under which passes an infinitely long tape marked with digital 0s and 1s. The

The Turing Machine: the imaginary forerunner of every modern computer. The machine shuffles the endless tape backwards and forwards, reading the symbols one by one and performing a logical operation on them. Alan Turing showed in 1935 that this simple machine, given enough time, can solve any 'computable' problem.

machine reads each bit in succession, either leaving it in place or replacing it with its opposite, in accordance with pre-received instructions. The Turing machine is a concept, a purely imaginary device, but it bears a remarkable resemblance to the tape-fed computers which were developed in the 1960s and 1970s.

Turing showed that such a machine, if fed with a suitable sequence of 0s and 1s on the tape, could solve any problem capable of solution by the powers of logic – i.e., by any logical procedure or algorithm. The term 'computable' is now used to describe problems solvable in this way. A problem, he demonstrated, was either computable or not. If it was, a machine of this kind, given enough time and enough memory space on the tape, would eventually solve it. If it was not, no machine which worked by means of mechanical rules of logic – not just this machine, but *any* machine – could ever solve it.

At the heart of a computer is a set of processing elements which manipulate binary logical propositions: truths and falsehoods and yeses and noes, represented by 0s and 1s. These are the sorts of logical manipulation at which computers excel and people are slow. Logical propositions are manipulated in hardware by passing them through elemental processing devices, 'gates' – so-called because they will 'let through' a 1 or a 0 if the answer is right and not if it is wrong. An AND gate has two input feeds, and will put out a 1 only if both the first and second input feeds are 1. An OR gate, by contrast, will let through a 1 if either one or the other input feed is a 1. A NOT gate, which has only one input, puts out a yes if the input is no, and vice versa; in other words, it reverses meaning. A few other gate types exist; for example, NAND, which combines NOT and AND by doing the AND and then reversing the result. Every computable problem can eventually be solved (as Alan Turing proved) by combining and repeating to a sometimes vast number of iterations these simple logical operations.

Turing went on to develop important ideas about computing. He and the Hungarian-born mathematician John Von Neumann, working at Princeton University, laid the groundwork for the idea that a logic machine should take the form of a collection of simple but fast gates plus a very large capacity of digital storage.

The story of the computer is one of the creation in silicon of ever faster and more powerful machines of this kind.

The story of Prince Louis de Broglie speaks well for the aristocracy. Born in 1892, he rejected the lifestyle which might have been expected of a scion of France's pre-Revolutionary monarchy and devoted himself instead to the problems of particle physics. He enrolled for studies at the Sorbonne, and in his doctoral dissertation produced an insight which made him one of the founding fathers of the twentieth-century revolution in science known as Quantum Theory. Society rewarded him with a Nobel Prize and Nature with long life. He lived until 1987, long enough to see the universal application of the extraordinary electronic devices which this new branch of science had made possible.

A present-day microprocessor chip performs millions of logical operations per second. The mystery is not that it can perform so many but that a completely solid piece of silicon in which there are no moving parts can 'perform' anything at all. If it were a lump of coal or iron we would expect it to . . . well, just sit there.

What is it, inside the chip, that is doing the performing? Although there are no moving parts in this small machine there are moving *particles*, tiny carriers of electric charge called electrons. Etched into the slice of silicon is an intricate pattern of paths they can travel. The paths, too, consist of silicon, but they are impregnated with minute quantities of impurity. The impurities change radically the conducting properties of the chip, guiding the electrons along well-defined routes and allowing them to interact in ways that imitate logical operations.

As the word suggests, electronics is about electrons. To understand what goes on in a silicon chip requires a grasp of the modern scientific view of these elusive subatomic particles. That view is given by what the great US physicist Richard Feynman used to call the 'strangely weird and beautiful' theory of quantum mechanics. Dr Chris Llewellyn-Smith, director-general of the international atomic research facility CERN in Geneva, put it another way: 'The electronics industry should be called the quantum physics industry.'

Unlike Newton's mechanics or Einstein's Relativity, Quantum Theory is not the creation of a single towering figure. In 1900 the German physicist Max Planck had the idea which gave the new science its name: that electro-magnetic oscillations are more like particles than we had imagined, in that they can exist only in little packets, or 'quanta'. This idea was further developed by de Broglie, who proposed – in his doctoral dissertation – that all particles of matter behave to some extent as waves. Before the development of Quantum Theory, science divided physical phenomena into two separate realms: the tangible world of mass and gravity, as described by Newton's laws, and the intangible world of charges, waves and radiation, as described by the electromagnetic theory.

At the turn of the twentieth century, however, science looked more deeply than had been possible before into the subatomic structure of matter, at the particles that made up atoms. With Planck demonstrating that electromagnetic waves come in particle-like packages and de Broglie showing that particles have the wavelike properties of frequency and wavelength, it was becoming apparent that nineteenth-century science's hard-and-fast distinction between particles and waves was disappearing. Every object, it transpired, is part wave and part particle. A hallmark of an object's particle-like quality is that it has mass and momentum, and a hallmark of its wavelike quality is that it will diffract (scatter around obstacles, as the signal from a cellular telephone does). Large everyday objects – people, for example – are essentially entirely particle-like: our wavelike characteristics are immeasurably small . . . when we pass through a door we hardly diffract at all. However, the subatomic components of matter – nuclei, electrons, photons and the like – certainly display combinations of both sets of characteristics.

Thus light can be described either as a wave of electromagnetic radiation or as a stream of particles (called photons); an electric current can be seen either as a movement of charged objects or as a movement of bundles of wavelike energy (called, either way, electrons). Each such particle or wave consists of a combination, in different proportions, of mass, energy, frequency and wavelength. Working out how these proportions look mathematically is a formidable problem; it was solved during the 1920s by theoretical physicists from Germany, Austria and Britain – Germany's Werner Heisenberg, Austria's Erwin Schrödinger and Britain's Paul Dirac.

It was Heisenberg who put together what has perhaps become quantum science's most famous idea, and the one which has thrown philosophers into confusion as to whether our knowledge of the world stands on as solid a footing as the confident nineteenth century had seemed to suggest: the Uncertainty Principle. This states that the mathematical characteristics of Planck's and de Broglie's particle-wave objects cannot all be known with certainty at the same time – that there is an intrinsic uncertainty, or probabilistic aspect, about their position, behaviour and mass.

Readers who do not understand this new conception of matter are in good company. Scientists and non-scientists alike are not fully comfortable with it. Einstein critiqued it with a famous assertion: 'God does not play dice.' Richard Feynman, himself a later contributor to the field, used to emphasize in his lectures that he did not understand it – which does not mean that he couldn't follow the mathematical models: what he was saying was that, in such a description of Nature, we have to reassess what we mean by the term 'understanding'. The indeterminate descriptions of Quantum Theory do not give us the comfort that things are *explained*. To this day philosophers are trying to come to terms with the change in our conception of knowledge which has been introduced by the Uncertainty Principle, which lies at the heart of Quantum Theory.

Whatever their philosophical ramifications, the new ideas constituted a revolution in science and shortly thereafter produced another technology: the technology of semiconductors, of modern electronics.

The familiar solar-system model of the atom (a central nucleus around which electrons move in orbits of varying sizes) predates the fully fledged Quantum Theory. Put forward by the English physicist Ernest Rutherford in 1911, this is still the best-known representation of atomic structure. However, it contains no explanation as to why the orbits should be where they are: it is like Kepler's picture of the actual Solar System – inspired and correct in placing the planets in elliptical orbits around the sun but, lacking any knowledge of the soon-forthcoming Newtonian laws of motion and gravity, powerless to explain *why* they were so.

Quantum mechanics provides the requisite explanation for the behaviour and internal structure of the atom. It supplies the laws of motion, so to speak, for electrons and for the other particles present in Rutherford's idealized solar-system atom. In 1913 the Danish physicist Niels Bohr applied the new science to show how the orbital positions of the electrons arise. Each electron, it turns out, has a particular, *quantized*, energy level. If it is at a low energy level it will be trapped in orbit around the atom's nucleus, of no use as a carrier of electric current; but if it has a higher energy it becomes a free electron, floating through the material and able to carry electric charge.

Metals like copper and aluminium conduct electricity because they contain plentiful free electrons. Insulators – for example, the plastics – do not conduct electricity because each electron is trapped in orbit around its individual nucleus.

In a third category are the semiconductors: these can conduct more or less electricity depending on how they are impregnated with small quantities of impurities. The most important of the semiconductors is silicon.

In the 1940s William Shockley was the leader of a group of scientists at Bell Laboratories studying the properties of this interesting group of materials. The object of their researches was to produce an electronic gate: a device which would let a current flowing along one path control the amount of flow in another. Shockley's invention, the transistor, was announced to the public in 1948. The transistor is the precursor of today's logic chips – a single gate within a fragment of silicon (or other semiconducting element: germanium was the one actually used in the earlier versions). The invention of the transistor was the definitive scientific achievement of modern electronics, and it earned Shockley the Nobel Prize for Physics.

Atoms of silicon are arranged in a diamond-like lattice, each having exactly four neighbours. Each atom also has exactly four electrons in its outer shell, each trapped in a bond with one of the neighbouring atoms. Those traps are where the quantized energy levels put the electrons – they are not free to roam, and so the silicon does not conduct electricity. It can be *made* to conduct, however, by the addition of minute quantities of an element that provides a supply of free electrons.

The additive used is phosphorus. Phosphorus is very similar in its atomic structure to silicon – it lies just one column to the right of silicon in the Periodic Table of the Elements familiar to high-school students of chemistry – but, in addition to the four electrons trapped in the bonds between the atoms, phosphorus has a fifth. According to the calculations of Quantum Theory, this fifth electron is at an energy level about one volt higher than that of the trapped ones. In a flashlight or a toy, one volt extra may not seem much, but in atomic structures each such quantum of energy makes all the difference. It is sufficient for the electron to be able to break free of its original atom in the lattice structure and roam freely through the material.

Silicon infiltrated ('doped') with an impurity such as phosphorus becomes a conductor. Unlike a metal, every atom of which has free electrons, doped silicon is not a very good one – hence its status as a semiconductor. Nevertheless it conducts much better than does an insulator, in which all the electrons are trapped around the atoms; moreover, by intelligent application of doping, the chip designer can insert conducting paths into the pure silicon.

In any semiconductor device, from the simplest transistor to a million-component chip, conduction paths of impurity must be laid where the current is to flow. In a modern microprocessor the conduction paths are exquisitely intricate. The clue to chip design is to channel electrons to follow paths through the silicon in such a way that they can perform logical operations. Interspersed along the conduction path are transistors, tiny silicon gates which allow electricity flow in one path to stop the flow in another.

Transistors are at the heart of any silicon device.

A transistor gate is made by introducing into the conduction path a region made up not of phosphorus, whose atoms have five electrons in the outer shell, but of another impurity, boron, whose atoms have only three. Instead of providing additional free electrons which can roam through the material, boron's atoms provide 'holes' – gaps in the bond structure which serve as additional traps for any wandering electrons.

At the gate the electrons from the conducting path fall into the gaps in the lattice – into the holes – and get trapped in the bonds. They are no longer free to carry current. A *depletion zone* develops – a region depleted of free electrons that therefore hinders the passage of current. The gate is closed.

The gate can be opened again by applying a voltage to it.

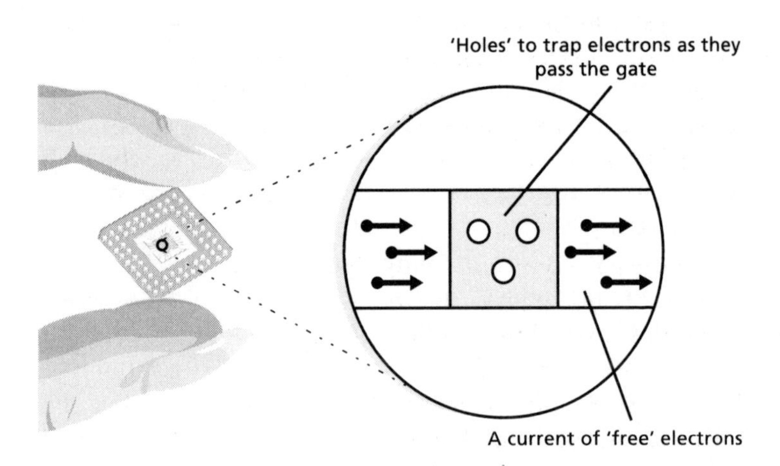

The silent machine with no moving parts. Inside a chip tiny gates open and shut, either letting electrons through or trapping them in 'holes' in the silicon.

By removing the negative charge on the gate – about one volt will do – the barrier is removed. As if the sluice in a canal lock had been opened, now the current of electrons can once again flow along the channel. The higher the voltage applied to the gate, the better will be the flow of current.

It is as true within a tiny chip of semiconductor as in the long-distance links of the telephone industry that a device cannot be digital – on/off – in strictly physical terms. Like all else in Nature, it does not have a definitive on or off. A transistor gate does not flick on neatly but opens and closes progressively as more or less voltage is applied.

The transistor was developed at Bell Laboratories with the objective of providing a means to amplify voice signals (obviously at that time analog), particularly to help the hard of hearing. In the years after its invention its best-known use was in the cheap transistor radios which flooded the world from the 1960s onwards. The new semiconductor device replaced the cumbersome vacuum-tube valve as the means by which the weak electronic signals gathered from the ether were beefed up to make audible sounds in a speaker. Wireless sets and gramophone players, previously items of furniture connected to the mains electricity grid, became pocket-sized and battery-powered; peace and tranquillity vanished from beaches.

Computer companies saw that the digital electronics industry could equally be transformed by this new invention. Provided the designer agreed to make it operate at two well-separated different voltage levels and call the higher one *on* and the lower one *off*, the operation became *effectively* digital; the analog way in which it actually switched – not quite instantaneously, and not precisely to the right level – would not matter. It was of course a waste of analog capabilities, of the ability of the transistor in a pocket radio to reproduce reasonably faithfully the varying levels of electrical signal it received from the airwaves, but even that could be turned to advantage; with the need to distinguish between only two voltage levels, particularly small and cheap transistors would suffice.

At the beginning of the 1960s IBM and other manufacturers produced the first large transistorized computers. These so-called second-generation machines (the first generation, marketed since 1951, had been vacuum-tube based) launched the computer age in its modern form. By today's standards these early second-generation machines were extraordinarily bulky; a typical corporate installation was housed in a specially built suite of air-conditioned and dust-sealed rooms. Even programmers stayed outside the room in which the actual processor and memory banks whirred; only a select priesthood of operators, dressed in white, could pass into this inner sanctum.

For the coming years these complex computing machines would embody and symbolize progress in the information age, but in a clandestine way; few people actually saw, let alone touched, them.

The next round of electronic technology would lead the computer out of the air-conditioned suite and onto the desktop.

After creating the germanium transistor Shockley left Bell to found his own semiconductor laboratory. In his case the transition from revolutionary inventor to sceptical boss took eight years. In 1957 two young scientists working for him, Robert Noyce and Gordon Moore, wanted to pursue research into another semiconducting material which had come into vogue, silicon. They thought it would enable them to create numerous transistors on a single wafer. Shockley refused, and they left to form Fairchild Semiconductor. A year later Robert Noyce, Fairchild's head of research, announced the invention of the integrated circuit. Moore was to become founder and chairman of Intel Corporation and a paragon of Silicon Valley entrepreneurship – scientist turned billionaire turned technology guru.

The concept of the integrated circuit is straightforward. If a single transistor can be made on a small slice of silicon, many transistors can be accommodated on a larger one. The components on the chip are insulated from each other partly by silicon's intrinsically poor conduction properties and additionally by oxidizing selected layers of the wafer at high temperatures – silicon oxide is an excellent insulator. The leads, resistors and other components which link up the transistors in a conventionally wired-up circuit are laid into the same wafer of silicon, also by means of chemical treatment. Electrical connection between the components etched into a chip is achieved by depositing onto the surface very thin leads of aluminium. The complicated matrix-like pattern of these aluminium leads gives the surface of the chip its awesomely intricate appearance – although in truth even that pattern tells only part of the story, the layers of doping underneath the leads adding another level of complexity to the circuit.

Implementation of this simple concept is, by contrast, far from straightforward. All the action takes place within a very thin layer, just microns (thousandths of a millimetre) deep, at the surface of the chip – the rest of the silicon is just bulk to make the chip physically less vulnerable. This layer contains in turn many even finer layers of doped regions (to produce the semiconducting effect), of oxidized regions (to improve insulation) and of metallic coatings (to improve conduction). Each of these layers is not only about a micron in thickness but contains in the horizontal plane patterns of paths and gates that themselves measure only a few microns across.

A chip designer is working at the frontier of what is possible. If the image of a chip is one of utter precision, the reality is different. The imprecision starts with the electrons themselves. It would not be right to think of the free electrons – or the trapped ones, for that matter – as little billiard balls making their way through gaps between the atoms. In the quantum world particles are not solid objects which can bump into each other, like the falling apple which inspired Newton's physics. They are de Broglie's wave-particles, better visualized as diffuse clouds of electric charge which drift though the atomic lattice. Their location and speed at any particular moment are, in accordance to the Uncertainty Principle, slightly vague.

Even when using individual transistors and wiring them together on a circuit board, care has to be taken to calculate what values the components of the circuit must have in order that the currents and voltages at the gates are in the correct ranges. The problem is magnified hugely when the wiring, insulating and resisting components consist themselves of doped silicon. Nowadays the designer makes use of very sophisticated computer software to help model the flows of current around the chip; but the process is art as much as science, relying on intuition to help find the patterns of doping which will achieve the desired interactions of electron flows.

By the late 1960s hundreds of transistors and sister components could be packed onto a single piece of silicon, and computer manufacturers were able to install a great deal more processing power and storage capacity into their machines. Physically, these got smaller; there appeared the minicomputer, which no longer needed a room to itself but could fit in with the office furniture. Minicomputer price tags were far lower than the million-dollar-plus level of a conventional mainframe installation, and the machines were bought for use by individual departments of corporations, and by smaller companies.

The definitive step forward, which has made computers the personal tools they have become today, occurred in 1974: Intel Corporation produced a chip, the 8080, which came close to being a whole computer in itself. It contained 4,800 transistors, arranged to provide all the key logic functions of a computer's central processor. The Turing machine had become a single piece of silicon, a *microprocessor.* All that was needed to make it into a fully usable machine was memory – the Turing tape, now also available in microchip form – and the clutter to interface with the outside world: keyboard, screen, mains supply. Manufacturers raced to produce such configurations and sell them to a now eager public.

The Apple II microcomputer, launched in 1977, set the standard by which future personal computers would be judged. Desk-top computers, affordable by every business and the average family, and operated by responding to helpful mouse-driven prompts rather than obscure typed commands, are so familiar now that it is easy to forget what a revolutionary idea they constituted. Steve Jobs, co-founder of Apple Inc., was probably not exaggerating when, reflecting back on those times in a speech in Boston twenty years later, he made the following summing-up:

> When we shipped the Apple II, you had to think differently about computers. Computers were these things you saw in movies. They weren't these things you have on your desktop. It was a totally different computer working in a totally different way, using a totally different part of your brain. It opened up the computer world for a lot of people who thought differently.

The invention of the chip set in train the extraordinary cycle of falling costs and increasing performance which has been the hallmark of the digital age. The chip was a great breakthrough not of science (it earned Noyce and his co-inventor Gordon Moore no Nobel prizes) but of manufacturing engineering. The scientific accolades had gone a decade earlier to the development of the transistor, a sort of one-switch chip. The chip is the miniaturization and integration onto a single slice of silicon of previously separated transistors. It marks the transition from electronics to microelectronics, achieving first a thousandfold and then a millionfold reduction in size and unit cost.

The electronics industry has benefited to a spectacular degree from that curiously ubiquitous law of manufacturing technology known as the learning curve. In the early 1940s industrial economists studying the production of aircraft wings for wartime fighter production became aware of a systematic trend in the cost of production. As the quantity of aircraft wings produced increased, so the cost of producing each one fell. But this was not the familiar economy-of-scale effect, which relates the unit cost of production to the scale of *current* manufacture. What was influencing the unit cost was the total number of wings produced *since the day their manufacture had started.*

These wartime aircraft engineers had discovered the learning curve, the mathematical relationship which determines by how much human labour efficiency increases as experience with a given task accumulates. It was to become one of the most widely used, if least understood, terms in the business vocabulary.

As the cumulative experience of producing each given component of a fighter aircraft increased, so the engineers responsible were able to find numerous ways of cutting the labour time associated with the task: through changing the workshop layout, the tooling arrangements and so forth. The mathematical relationship discovered was the following: each time the cumulative output of a component doubled (cumulative meaning the total output since the first date of production), the cost of labour input fell by a constant factor – approximately 25 per cent.

Similar measurements were subsequently made, mainly by the Boston Consulting Group, of the relationships between unit costs and cumulative output in a very wide range of industries, from machine production to chemical plants, from industrial tools to domestic appliances. What seemed a cast-iron law of economics emerged. As the cumulative output of a manufacturing process doubled, the unit cost fell by 20–30 per cent. Technically, if the cost trend being traced is that of labour input only, the curve is known as a *learning curve*, while if overall input costs, including capital and materials, are measured the curve is labelled an *experience curve*. Either way, it is remarkably uniform and universal.

New generations of computers, display devices, memory systems and telecommunications technologies appear with such speed that the industry appears to be going through a fresh revolution every one or two years.

But the rate of development of electronic devices is actually less surprising than it appears. The microelectronic chip is on the one hand greatly complex and exquisitely miniaturized, but on the other hand it is of considerable conceptual simplicity. There are constraints on how far this technology can take us, and on how quickly it can take us there. When the accessories and frills are stripped away and the basic components studied, it transpires that in the last thirty-five years the key technology – microelectronics – has developed at a measured pace and in a well-defined direction.

As the chips have developed in speed and capacity, as the industry has turned out successively the 1K-, 4K-, 16K-, 64K- . . . through to the 256M-bit chip, it has done so with a regularity which has not taken experts by surprise. In 1965 Moore predicted that the number of components which could be packed onto a chip would double each twenty-four months. Moore's law, as it has come to be called, has governed the rate of development of electronic devices during the past three decades. The Intel 8080 chip, dating from 1974, had fewer than five thousand transistors. Two decades later, the Pentium II had over five million.

While Moore's law talks of how far the miniaturization process can be pushed, it is equations of economics (such as the learning curve) which govern cost trends. These have been extensively studied – hence the well-known quips to the effect that 'a Cadillac would now cost $5 if car prices had followed the trend of chip costs'. Lines on graph paper tracking the fall in cost per unit of performance – dollars per megabit of memory, dollars per MIP (Million Instructions Per Second) of processing power – show rapid and uniform cost declines throughout the past three decades.

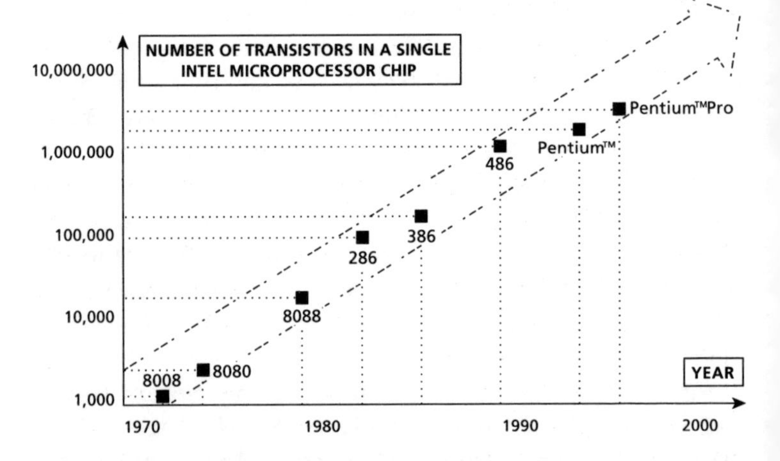

Gordon Moore, co-inventor of the chip and later founder of Intel, stated in 1965 that the number of transistors that can be crammed onto a single chip will double every 24 months. This relentless exponential trend, 'Moore's law', is the most important driver of progress in information technology.

★

Chips contain not just logic gates but memory – the tape on the Turing machine. The electronic storage device is a capacitor. A capacitor is made by deliberately placing two conducting pathways very close to one another, separated only by a thin slice of insulation. If a voltage is applied to the conductors no current of electrons will flow between them, because they are separated by the insulator. But electrical energy will build up between them – an area of electric charge is formed in the insulating layer, which acts as an energy store. (The configuration is called a capacitor because it has the capacity to store electrical energy in this way.) Each tiny area is a storage point of information; it is either electrically charged (representing a binary 1) or not (a binary 0). A modern RAM (Random-Access-Memory) chip has many millions of these storage points. They are a programmer's dream data store: millions of bits of instantly accessible memory, embedded in the silicon alongside the transistors which control access to it.

Other computer components, not just chips, are benefiting from the same trends of miniaturization and learning-curve economics. The really large areas of data storage in an information-technology system are rotating mechanical devices: hard disks, floppy disks, CD-ROMs. In a disk drive, a read–write head is guided over the rotating disk surface by a little robot arm. To record a bit of data a pulse of current is used to magnetize a spot on the rotating surface; to read off the data, the procedure reverses, the magnetized spot generating a current in the head. Based on a principle no more sophisticated than the vinyl gramophone record familiar to generations of music lovers, the technology of rotating mechanical memory is developing at an astonishing rate; like the chip, it benefits from the same trends of miniaturization and learning-curve economics.

The first hard drive sold by IBM in 1956 contained fifty

disks each of diameter 60 centimetres, required a truck to deliver it, cost over a million dollars at today's prices, and contained just five megabytes of data. Since then capacity per disk has roughly doubled every two years, a rate of progression which has actually been speeding up in the 1990s. A typical hard drive inside a home PC today stores thousands of megabytes of data on tiny disks just several centimetres across.

Storage density on magneto-resistive systems, as these are called, has reached about 150 megabytes per square centimetre. Researchers estimate that this could be increased to the range of 15 gigabytes per square centimetre before a limitation of basic physics – the tendency for thermal activity to change the magnetized state of individual particles randomly – begins to make further miniaturization impracticable without a change of storage medium.

While electrical engineers worked to perfect the design and fabrication of silicon chips, a new category of information-technology professionals came into being: programmers – called also by the more formal title 'software engineers'. They practised an odd kind of engineering, for they did not work with silicon or phosphorus or boron, or indeed with any of the elements which make up our material Universe at all, but with strings of 0s and 1s. These were the people who created the marks on the Turing tape, giving the machines the instructions they needed to solve the tasks they were set.

It was the perfect mental challenge for the young and technically minded, for those in their late teens and early twenties with a bent for the analytic and quantitative. There was no need for deep knowledge or extensive training in chemical engineering, aeronautics, fluid dynamics or even electronics. What was needed was the ability to cope with streams of symbols, those you wrote down and those the computer spewed out on reams of perforated printer-paper. A minimum of equipment was required: just paper, pencil and shared access to a computer (there were no desktop machines in those days). Because this was not 'real' engineering – a bridge would not collapse nor a train crash if you got the sums wrong, the computer would just print out nonsense – there was no need for years of apprenticeship and the acquisition of qualifications. A unique job culture developed, at once casual but intense, one of working odd hours (the middle of the night) in odd places (by the computer room to get quick turn-around on the many trials of each program before the code was finally worked), and wearing T-shirts and jeans rather than jackets and ties.

The earliest programmers, working in the 1950s and 1960s, did actually write in the language of 0s and 1s that the computers read off the tape and fed into the chips for processing – this was, and still is, called machine code, and requires an intimate knowledge of the workings of the hardware. They did not have the luxury of writing just 'Print payroll' when they wanted a printout of the payroll file. They had to specify in which precise spot in the computer's giant memory array each digit in each entry in the payroll file was stored, and then specify to the printer the exact spot on the page to which that digit should be posted. To achieve a neatly arrayed tabulation on a single sheet of printout could take several attempts and hours of work.

But, as the years went by, new tools came to the programmers' aid. Sophisticated programs called *compilers* were developed; the code-writer could enter a word like 'print' and this program would compose – compile – the hundreds of machine instructions which were required to get the printer to put the array of characters correctly on the page. Most of today's programmers, both professional and amateur, have never had to descend to the level of machine code, or even to be aware exactly where in the memories and processors the computer is storing and manipulating its binary numbers. They write in so-called *higher-level* languages, using words like 'sort' and 'copy'; the tedium of translating these into code that the machine's electronic devices can interpret is left to the compiler.

As we sit at our PCs today, working our way around the screen with mouse and pointer, we are 'programming' the machine to respond to our requests in a very high-level language indeed, one elevated practically to the level of everyday English. We click on words like 'help' and 'check spelling', blissfully oblivious to the need for our software to issue thousands of machine instructions before the task, moments later, is accomplished.

The second way in which the programmers' jobs were made manageable was the development of modular software design. The overall task – say, to produce a video game – was divided into parts (modules). One of the modules might be concerned just with producing the screen image of the gun-toting hero. The programmer does not have to redraw the hero each time he appears (which might take a month of full-time work if done from scratch) but just summons up a module which says, 'Hero appears on left of screen.' A modern software suite for a major application like operating a telephone exchange can run to millions of lines of code: no single human being is aware of the totality of the program, because it is divided into so many modules.

In the 1970s, as computers entered the workplace in increasing numbers, the applications became of huge commercial importance: entire financial, industrial and distribution operations relied on them for their smooth functioning. Module was piled upon module as software suites grew in size and complexity. Nevertheless, the rather unique culture of the programmer lived on; it was a professional world distant from much that was going on in the industries which employed these people. Strictly speaking the programmers were engineers, but their expertise was, and still is, streets away from that of civil, mechanical or electrical engineers. Mechanical stresses or chemical reactions are not their areas of concern, nor indeed is the flow of current in wires – many programmers are not trained in electronics, the very stuff of digital processing. The world in which the software experts of today feel most comfortable is cyberspace. They operate in the purest of information worlds: just bits and bytes, abstracted even from the cables, chips and radio waves via which the software will run.

Another concern which did not loom large for this young community of people writing code in the 1970s was the distant future (believe me, I was one of them). It was in this unrestrained and rather swashbuckling professional atmosphere that the seeds were sown for the AD 2000 'millennium bug' – Y2K, to use the jargon. Scarcity of computer memory has always been a headache to users. In the 1960s and 1970s it was in such short supply that every available technique was used to conserve it. One of the memory-saving habits was the use of only the last two digits of the year when working with calendar dates: 10 April 1978 would be 10.4.*78*, not 10.4.*1978*. Nowadays memory has become so cheap that is seems difficult to imagine that programmers would bother to save bits by lopping two digits off a date. But this was two decades ago: memory seemed very scarce and the year 2000 very distant.

The practice was to have a huge unintended consequence. As the year 2000 loomed, the computer community realized that many lines of code written in those earlier years were still running on machines throughout industry and commerce, often buried in modules that were in turn buried in larger modules within still larger suites of more recently written programming, suites so complicated that no one knew where the offending items might be hiding. At the turn of the millennium the programs would register the new year, 00 to the code, as 1900 instead of 2000, and the program would produce nonsensical results. The computer would act as if the world had jumped back to 1900.

Would that the computers in banks and factories and hospitals had a shred of common sense! They could glance through their digital security cameras, like ISAAC in the fictional Gridiron Building, and see that it was clearly not the time of Queen Victoria, Kaiser Wilhelm and President McKinley. And, anyway, if yesterday it was 1999 then today it can't be 1900, can it?

But computers don't have common sense. What they have instead is hundreds of thousands – even millions – of lines of programming. Most of this has been written recently, so the years are fully spelled out in four digits and 1999 gracefully makes way for 2000, but the programs are also riddled with much older code, long-forgotten modules written perhaps in the 1970s or early 1980s: the original author retired long ago, or has moved elsewhere, or has died, and all traces of the notes which he or she made to document the workings of the code are lost.

Computer programs keep checking the date. Not just in obvious situations, as when an accounting program needs to know what date it is before printing out an invoice. They do it on the slightest whim of their author. To take one of a thousand such examples, a (now retired) software writer might have programmed a fire-alarm system to check the calendar before sounding test alarms, so that the test would be postponed if the day were a Sunday. When the year 2000 bug throws the logic of the calendar, the system might accidentally run a test alarm on a Sunday and, getting no human response, close down the power, turn on the sprinklers and summon the fire brigade.

As for the programming suites in the financial sector, dates are integral to their every calculation. Time is money – the onset of a new day means interest is payable, options expire, and contract clauses are triggered. A calendar error of just a single day in one bank's software could trigger enough confusion to launch myriad lawsuits. A shift of 100 years in dates throughout the worldwide financial system and you could have, to use a word often employed in this context, meltdown.

Bank-interest calculations would be out by a ludicrous multiple as the software shifted the date by 100 years. Supermarket programs would think supplies were arriving a century too late and take nonsensical corrective action. Mostly, the programs would just 'crash': splutter out with their dying breath an ERROR message, then stop.

Preparations to combat this millennium bug were to cost the public and private sectors many billions of dollars in the last years of the 1990s – as the programming community, which had earned its living creating the problem, earned it afresh, and at emergency rates solving it.

The microelectronics industry has succeeded in doubling every two years the number of electrical switching points etched into a single silicon slice. Technical progress has reduced computers from the size of a car to the size of a notebook. The production of information-processing and -storage devices is the most spectacular instance yet of humanity's ability to improve manufacturing techniques. Although the learning-curve effect – the reduction of production cost by some 20–30 per cent with each doubling of output – is not special to this industry, the numbers have worked particularly well. The massive demand for these devices has ensured that the 'virtuous circle' of falling prices and improved performance leading to yet further demand and hence yet more performance continues. Since the devices are manipulating a weightless and dimensionless entity called information, there is the additional potential, unique to information technology, of almost limitless miniaturization.

There is a great deal of discussion as to how long the trend can continue. Ultimately physical limits *will* be reached in the number of components that can be packed into a chip. Logical operations are carried out in silicon by the movement of electrons along conducting paths in the chip. The finite dimensions of atoms and electrons – or, more correctly, of the areas over which their effects are felt – are such that, if the conducting paths get too narrow, the effects begin to interfere with each other. There is considerable room for further miniaturization, however, before these limits are reached: there are no signs that 'The Wall', as it is sometimes called, will be hit until there are well in excess of 100 million transistors in a microprocessor – and physics laboratories are full of ideas for going much further than that.

The emphasis for the next several years will be on manufacturing technique. The process of etching the various layers of doping in a chip is basically a photolithographical one, and the first barrier will be an inability to focus light rays sharply enough to make the components sufficiently small. Ideas include moving up to the high end of the ultraviolet spectrum, where wavelengths get very short and hence the sharpness of the beams doing the etching is improved. The next step would be to use X-rays.

Eventually the problem will shift from one of manufacturing precision to how tiny the paths in the silicon can actually be and still work reliably. Current is a flow of electrons, and electrons are quantum phenomena – we cannot be sure exactly where they are at any time. You need a reasonable number of electrons to be inside each gate for it to be reliably *on* or *off*. In the densest chips electrons are flowing into and out of components in bundles of around

10,000, still leaving much safety margin for any imprecision in their exact individual locations. But at the present rate of miniaturization it will be not many years before the numbers become worryingly small.

At that point the conventional transistor-as-gate design principle will have reached its limits; but scientists are looking ahead to this as a challenge rather than an obstacle. Why not make very exact control of the number of electrons the new design principle, ideally one electron per bit? You cannot get much more digital than that! Researchers at Cambridge University have indeed constructed an experimental device that stores one bit as one electron, and have received substantial funding from Hitachi to take the idea further.

With the development of the digital computer, technology has now succeeded in simulating the mental activity of information processing – the second of the three levels (communications, processing, thinking) at which we as humans manipulate information. To be more exact, it has achieved the ability to simulate one aspect of the way in which we process information: the solving of computational problems, those which can be reduced to the shuffling of bits on the tape of a Turing machine.

With the emergence of hardware and software of such astonishing complexity and problem-solving power as is available today, it is easy to forget that at the heart of every computer is nothing more than an array of AND, OR and NOT gates manipulating binary logical propositions. The gates in today's most advanced chips are performing exactly the same logical operations as did their vacuum-tube counterparts half a century ago, and they will be doing likewise half a century from now. Hopeful names are given to attempts to break the mould: 'artificial intelligence', suggesting mind-like thinking; 'neural computing', invoking the brain analogy still more strongly; or 'fuzzy logic', implying freedom, finally, from the cold determinism of digital processing. But all that is changing, in the chips themselves, is speed and capacity.

A Turing machine, however simple, can be programmed to solve any computable problem. More complicated versions, electronically powered to speed them up and working with many bits in parallel (today's 16-bit or 32-bit processors), will solve problems faster, but cannot solve any which the simple one cannot. Turing set out once and for all, before a single programmable computer had been built, the path which society's new technology project would take: to build ever better devices for processing digital 1s and 0s (faster and higher-capacity Turing machines), but knowing in advance that no machine ever built will answer a question that lies outside the 'computable' category.

Measured by changes in the power of chips, or growth in the capacity of networks, the digital revolution is without precedent in history. But the computer phenomenon has been launched along its remarkable path by a single extraordinary invention in 1958, that of the microelectronic circuit. Since then there has been more evolution and less revolution than meets the eye. Like the development of mechanical power at the outset of the Industrial Revolution, the invention of an electronic device to process data has basically been 'done'; that device is now working its way into the corners of our workplaces and homes.

5

But Are Computers Like Us? The Rise and Fall of Artificial Intelligence

In May 1997 the IBM computer Deep Blue played against world champion Grandmaster Kasparov at chess. For a week the computer's progress made front-page headlines; the match was the most widely reported in the history of that ancient game. When Kasparov lost, there was a sense that a milestone had been reached. The computer, it seemed, had finally triumphed in a contest that pitted it against human powers of thought.

There was much one could say in defence of the human. The game has millions of possible combinations of moves, making it well suited to computational analysis; more importantly, the computer was designed and programmed by people, so the match was really a triumph of the (human) designers and programmers, not the machine per se. Nevertheless, the sense that the digital age had produced something which had been taught to *think* was palpable. In the words of chess Grandmaster Yasser Seirawan, 'What on earth are we doing when we play a game of chess if not thinking?'

If a machine can be taught to beat the world's best practitioner in one of humanity's most intellectually demanding professions, surely it will be able to replace people in other occupations: in marketing, stockbroking, medicine or the law? Or, if not replace them, provide them with artificial-intelligence software which makes each of them the equal of the finest minds in their field? The implications for the value of raw, unenhanced, human thinking seem immense.

Technology had conquered the first level of the communication–processing–thinking hierarchy with the construction of the telecommunications infrastructure of cables, fibres and radio waves. It had cracked the second level with the introduction of high-speed digital computers. Now the challenge which remained was to reach the third level: to make a thinking machine.

In TV cop serials there is a lot of emotion. Subordinates shout at police chiefs and police chiefs shout back. There is swearing and the slamming of doors. Officers experience moments of elation and of dejection, follow hunches, find clues in a gesture or a look in the eye. And in the end they solve the crime. In a real police force there is more boredom and bureaucracy, and the crime-solving rate is lower. But the fact remains that, in reality as in fiction, the character of the working environment is set overwhelmingly by interaction among people.

The same is true in business. Regardless of how well structured the working routines of an employee in a corporate office might be, billions of brain-cells are gnawing away at age-old interpersonal issues: Why is the boss looking at me like that? Why do I always seem to be working harder than Joe? What can I do to get a better job here – or an even better one in the new company down the road?

Computer-phobes and computer-enthusiasts alike recognize that, before they can make any further dramatic inroads into human organizations, computers will have to develop more human personalities. They will need to have moods, humours, intuitions, maybe even occasional fears – otherwise people will not be willing to deal with the machines and so the machines will not be able to deal with people. Computer technology will have to process information not in the narrow sense of manipulating symbols in accordance with the strict logic of a Turing machine but in the broader sense in which the human mind builds up knowledge.

In short, computer intelligence has to become more human. And there are researchers in our best universities and leading research laboratories who believe we are going to achieve this.

The idea of artificial intelligence (AI), of a machine that imitates the human power of thought, goes back to the earliest days of computing. The first AI software was conceived in 1955 by Herbert Simon, Professor of Psychology and Computer Science at Carnegie Mellon University and future Nobel Economics Prize winner. With fellow computer scientist Alan Newell he wrote a program designed to produce proofs for simple mathematical statements. Dubbing it the Logic Theorist, he announced to a startled classroom of students that he had invented a 'thinking machine'. That this first application of a machine to problem-solving was in the field of mathematics is perhaps unsurprising; the connection between computers and precise logical analysis is an enduring characteristic, which was to become both the computer's great strength but also its key limitation.

It was Turing who had constructed the framework for this debate, as for so much else in the computer field. In 1950, a decade and a half after he had conceived of his imaginary tape-fed machine, he published a paper under the title 'Computing Machinery and Intelligence' in which he proposed that the question 'Can a machine think in the manner of a human?' could best be tested by seeing whether it could *interact* with people as if it were indeed one of us. In other words, the criterion Turing adopted, which had great merit from both the philosophical and the practical viewpoint, was the ability to 'imitate people'. What he called the Imitation Game – it is now termed the Turing test – involves placing interrogators at a computer terminal connected through a wall to either another human or to the machine which is to undergo the test (the questioner does not know which). If the interrogator mistakes the computer-generated replies for those of a human, the machine has passed the test.

In the mid-1960s Arthur C. Clarke, together with the film director Stanley Kubrick, wrote the script for *2001: A Space Odyssey*. In it the computer HAL has evidently human qualities – not only intelligence but emotion and self-awareness. Clarke's earlier forecast in the field of high technology, his prediction back in 1945 that we would be bouncing messages off satellites in space, had already come true. This forecast of human-like computers by the start of the new millennium seemed hardly less credible, and by the 1970s 'AI labs' were springing up in research institutes and universities. AI had become the hottest buzzword in the computer industry.

It was recognized that the highly sequential logic of existing digital machines was a long way from the processing operations performed by the biological brain; therefore electronics had to escape from the shackles of the linear computational model. Various concepts, inspired by the structure of the brain, were brought forward to inch computer scientists towards their new goal.

Microscopic examination of brain-cells by neuroscientists revealed that, in learning, a human mind goes through two distinct processes. As we engage in thinking, our neurons fire signals to each other along tree-like axons and dendrites which are the wiring between the neurons. But those connections themselves evolve over time; they get stronger as a given activity is repeated, so that the activity gets 'coded in' at a deeper level than the conscious. The ability of the brain to learn through the strengthening of these connections – synapses – is its *plasticity*, and is most evident in childhood. The ability to use the network once the connections are substantially in place is the brain's *power*, and is at its best in adulthood. The difference between plasticity and power can be seen in the difference between the way children and adults learn a language. When Wim Duisenberg, the Dutch head of the new European Central Bank, appears on US television to explain monetary policy, he produces complex English sentences with a grammatical craftsmanship that a kid from the Bronx could not match. But when the Bronx kid speaks it is with effortless fluency and complete naturalness. The slightly stilted precision of the Dutchman, by contrast, reveals that the language he is using is foreign to him, and that he needs to use his analytic powers to put those sentences together. Duisenberg has learnt to speak English using the *power* of the brain, the kid from the Bronx using its *plasticity*.

The 1994 German Business Innovation Prize was awarded to Siemens Nixdorf for a machine they termed the Synapse-I Neural Computer. The manufacturer, a European corporation not normally given to verbal extravagance, advertised it as a 'thinking' computer, which could learn and be taught. It was described as capable of tasks such as medical analysis, control of complex industrial processes and 'predicting interest-rate movements more accurately than financial gurus'.

Neural-network machines, of which the Synapse-I is an example, are the latest manifestation of a trend which began in the early 1980s, following pioneering work at the California Institute of Technology. Instead of designing a computer around a single very fast and sophisticated central processor, a large number of tiny processing elements were interconnected in parallel in a manner reminiscent of the synapses which interconnect neurons within the brain. Companies with names like Thinking Machines Corporation produced hardware dedicated to running the new programs – hardware whose many processors, running in parallel, allowed the simultaneous processing of many different 'lines of reasoning' before converging on solutions.

The name Siemens Nixdorf chose for their machine is revealing. Neurons, synapses – the anatomy of the brain was being invoked ever more explicitly to inspire the design of computers. In neural computer networks, the connections between the logic elements are made to vary in strength in a haphazard way, reminiscent of the variations in the strengths of neural signals travelling across biological synapses. The computer is taught a task repetitively and, as it learns, the strength of the 'synaptic' connections gradually adjusts until the task is more or less mastered. In other words, the computer is not rigidly programmed from the start in accordance with a logically determined formula, but gradually evolves a set of connections between the many individual logic elements – rather in the way that a child, learning to ride a bicycle, unconsciously adjusts the wiring of the brain to reflect the new skills being learnt. The programmer does not know how the computer is doing the learning; he or she just makes sure the machine keeps trying and can recognize success from failure.

Companies on Wall Street use such techniques to spot trends in stock prices from the avalanches of data which the market produces every trading session. Neural nets, as they are called in shorthand, are particularly good at picking out patterns in data – trends or correlations which are difficult to trace using traditional mathematical tools because you may not know what kind of pattern you are looking for. The use of neural-network techniques by MasterCard International to identify patterns of credit-card fraud has prevented losses estimated at $30 million annually.

A further twist on the biological analogy was inspired by the theory of Darwinian evolution, and was called the genetic algorithm. The idea here is not to write software in its final form but to let it emerge by trial and error. Little snippets of code (called chromosomes – another biological borrowing) are thrown together haphazardly in numerous combinations and put into an environment where only the 'fittest' – those best able to reproduce – survive. The program is written in such a way as to allow the chromosomes to duplicate themselves. Random mutations are thrown in to ensure, as in biological evolution, that there is the potential for progress. Stock-market applications of the genetic algorithm are already gaining acceptance, the software offering clues to trends that even neural-network technology could not find.

AI enthusiasts have seen expert systems, neural nets and genetic algorithms as incremental steps – each blessed by acceptance in the tough world of commerce – towards the ultimate goal of a machine which exhibits mental powers which are a challenge to our own.

To MIT's Marvin Minsky – one of the outstanding figures in the AI community – although much of its microstructure may still be a mystery, the brain is nevertheless just a machine in which electrical and chemical signals interact and are stored. In due course, he believes, we will be able to understand its structure and copy it – or, at least, copy it sufficiently closely to produce machines which reason in broadly human ways.

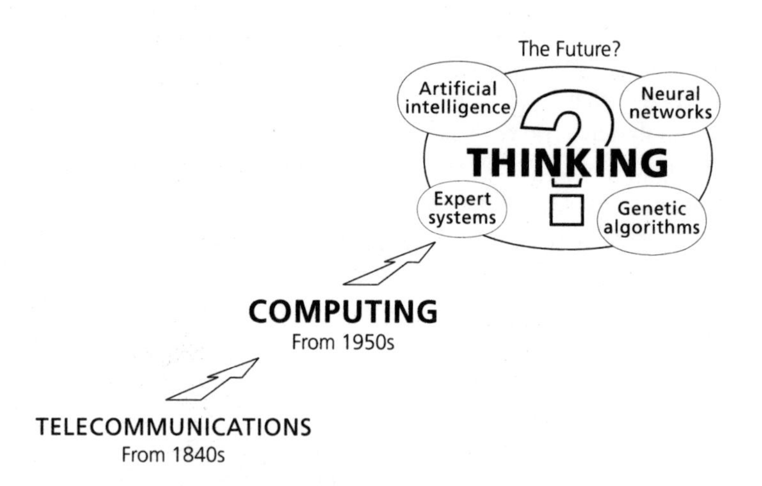

Communication, processing, and then . . . a thinking machine? The attainment of this goal faces philosophical and practical obstacles which are, in the majority view, insurmountable.

Yet there is a long-running sense among philosophers and scientists – and many of the rest of us – that there is more to thinking than computing. The idea that our minds do in some sense transcend the confines of purely logical reason is intuitively appealing, yet this has been difficult to pin down. Philosophy has long recognized that there is a deep-rooted difference between syntax (symbol manipulation) and semantics (meaning); translating this to the information-technology debate, we would say that, though computers manipulate *symbols*, this is not the same as giving them *meaning*. The arguments against the possibility of AI address, in different ways, the question of whether there is more to thinking than the shuffling of symbols in accordance with mechanical or algorithmic rules.

A watershed in this debate came in 1931 with the publication of the famous Incompleteness Theorem by the

Austrian logician Kurt Gödel. His contemporary Alan Turing had shown, with his imaginary machine, that, given enough time and memory, a device with simple gate circuits could solve any problem amenable to logical solution – any problem which could be solved by breaking it down into elementary logical steps and then reassembling the whole. Gödel, like Turing, was interested in exploring the boundaries of the power of logical reasoning. He discovered that, if you start with a set of (logically consistent) premises which lead to certain consequences or conclusions, then the powers of logic will be unable to lead you from those premises to *all* of the consequences or conclusions.

The intuitively appealing idea that our minds can in some sense transcend the confines of purely logical reason has been given a great boost by Gödel's Incompleteness Theorem. In simplified terms, he showed that not even all logical problems (let alone those which are not framed in logical terms) are capable of being logically solved. In Turing's terminology, not all problems are computable.

The finding dealt a great blow to a hope which had been growing among philosophers that there was a perfection about logic, something that could really make it the 'gold standard' for human reasoning. It was a hope that could be traced back to Pythagoras and Plato, but it was particularly energizing minds at the start of the twentieth century, most notably that of the British philosopher Bertrand Russell. Russell was bothered that there seemed to be holes in the logical enterprise, specifically that you could make statements which were entirely correct in their formulation but turned out to be logical paradoxes. For example:

> There is a barber in a village who shaves all (and only) the men in that village who do not shave themselves.

It sounds a perfectly sensible sentence – indeed, a sensible state of affairs: each man either shaves himself or goes to the barber. Unfortunately a paradox emerges when you ask whether *the barber* shaves himself or not. A minute's analysis of the sentence reveals that if he does then he doesn't, and if he doesn't then he does. The statement is, strictly speaking, non-sense – or nonsense. The barber paradox, rephrased in the formal language of logic, addresses the question:

> Can there be a set of all sets which are not members of themselves?

The answer is that there is no answer. The statement is equally illogical in this more mathematical form.

Traps like this worried Russell greatly. If logical reasoning was to work it was important to have a set of axioms, or starting points, which prevented you from falling into illogical traps like that concerning the village barber.

With a colleague, Alfred North Whitehead, Russell set about writing the *Principia Mathematica* (1910–13), their goal being to set out a complete framework of logical foundations which did not fall into this trap. It was a magnificent enterprise, but it failed. When Gödel's Theorem was published the world knew (as did Russell; he lived until 1970) that the enterprise *could never* have succeeded. There is no such thing as a full and consistent system of logic: a logical system is incapable of proving its own premises without drawing from outside itself.

Gödel's Theorem has become a turning point in our thinking about logical reasoning. It has done for mathematics and logic what Quantum Theory has done for physics – reintroduced an element of fuzziness into fields which had seemed to be getting ever more clear. The Gödel–Turing distinction, between problems which are solvable by what we might describe as engineering products and those which are not, is one of the foundations underlying the argument that computers will never replicate human thinking. On this basis the Oxford philosopher John Lucas was able in the 1960s to develop his influential arguments against the whole AI concept of the ability of computers to think. There are arithmetical truths which (in the words of mathematician John Casti of the Santa Fe Institute) 'we humans can see to be true but a machine cannot prove'.

There is an old schoolboy joke:

Teacher: Jonathan, what is two plus two?

Jonathan: Four.

Teacher: Good.

Jonathan: What do you mean, *good*? That's perfect!

The teacher is exasperated, but the fact is that Jonathan has the right of the argument. The outcome of the logical manipulation of digital entities is not good or bad, better or worse: it is perfect (right) or it is wrong. (The teacher will have to watch out for this pupil, because he has grasped at a very early age the difference between digital and analog thinking!)

Sounds, by contrast, are analog, so our teacher would be on firmer ground here. A French lesson:

Teacher: Jonathan, pronounce the word Monsieur.

Jonathan: M'sieur.

Teacher: Good.

This time Jonathan can't snap back 'That's perfect!' because there's no such thing as a perfect way to pronounce a word. One can describe the way a word is said as crisp or slurred, melodic or flat, tense or relaxed, emphatic or unassuming – even beautiful or ugly – but not as *perfect*. (Digital computers experience, to this day, considerable difficulty in recognizing spoken – analog – words.)

Some years later, a history lesson:

Teacher: What were the causes of World War One?

Jonathan: Rising military tensions between the major European powers from about 1910, sparked off by the assassination of Archduke Franz Ferdinand in Sarajevo in June 1914.

Teacher: That is certainly one of the acknowledged interpretations, but can you expand on those rising tensions?

As Jonathan grows older and takes lessons in history, literature, economics and business studies, he will wander into territory further and further from the logical. He will find that there are two kinds of problems: those in which answers are demonstrably right or wrong, and those which permit many interpretations.

Logical operations – call them digital operations – produce restatements of facts which are already present before the operation begins. Two plus two equals four because four is another way of saying two plus two. One of the problems set in a question in the 1970s to students applying for entrance to study mathematics at Cambridge University was to *prove* that two plus two equals four. It was a surprisingly tough question for school-age students. The clue to the answer is the definition of each whole number (integer), which can be found in any dictionary: each integer has value one more than the integer preceding it. 'Two' thus means 'one more than one', 'three' means 'one more than two' and 'four' means 'one more than three'. With this realized, the proof is easy enough. 'Four' means 'three plus one', which means 'two plus one plus one' which means 'one plus one plus one plus one'. 'Four' and 'two plus two' are, therefore, both ways of saying 'four ones'.

What could be called a law of conservation of information applies to all logical operations: information is neither created nor destroyed. This single insight sets apart those tasks which logical thinking can address and those which it cannot. Logical thinking can address those problems in which the answer is already contained in the definition of the words which make up the question.

A small part of the processing which humans undertake is of the strictly logical – digital – variety: it applies black-and-white rules such as those of arithmetic or formal logic. Here computing machines, which after fifty years are still very much digital, are incomparably superior in their speed of operation to any human mind. Digital computing, to rephrase, is about the application of artificial machinery to those aspects of human thinking in which the manipulation of mental variables is logical, discrete and precise. But the full gamut of human processing of information is supremely complex and multifaceted; and most of it is analog, as in the very important areas of visual and sound recognition. In his book *What Will Be* (1997), MIT's Michael Dertouzos writes extensively about the future of the new technologies. By and large he is optimistic about technology, but cautions that:

> Show me a computer that can look at a video and identify that a little girl with a white dog on a leash is running across a street and is about to get run over by a BMW convertible, and I'll show you a fake!

In the manipulation of numbers and alphabetic text, computers achieved very considerable success early on. Technology could calculate faster than the human mind even in the days of the first digital machines, and word-processing programs had achieved a high level of sophistication by the 1970s. It would be understandable if one expected similar progress in the interpretation of video, audio and other signals once methods of digitizing them had been found.

But this is to forget that the ability of machines to manipulate quantities and words digitally was not attributable to fifty years' worth of advances in computer technology but to the development, over the preceding five *millennia*, of their digital forms (10 numerals and 26 letters of the alphabet) and of logical rules for manipulating them (the rules of mathematics and logic). In the case of video images, the computer fraternity has had to start from scratch – as if they were trying to work with numbers before the rules of arithmetic had been discovered or with words before the days of alphabets and grammar.

As of now the world's best laboratory software cannot derive from a photograph of a room a simple description of its contents: chairs, tables, curtains, pets, people. And even when software does manage to acquire this apparently simple ability – give the boffins some more time – it will still have no idea of the *significance* of the presence of the objects. What do the layout of the chairs, the design of the carpet and the breed of the dog say about the owner? The picture will not be *understood* by the machine. What the analog–digital conversion process can do with great precision is to identify patterns of light or sound. The computer can scan the pictures and sounds, even recognize them provided it is given extremely clear guidelines as to what it is looking for, but what it can't capture is what the pattern *means*, even at levels we would consider, if applied to human beings, very rudimentary.

With the advent of digital technology machines have crossed the boundary between levels one and two of the data–information–knowledge hierarchy: in addition to transmitting data they can recognize and process its information content. But the processing is limited: the information can be recognized but it cannot be understood.

The difference between the computable and the non-computable is the same as that between the manipulation of information using algorithms (rules, procedures, instructions) and non-algorithmic processing. Philosophers and mathematicians versed in the writings of Gödel talk in terms of computable versus noncomputable. The layperson can quite correctly talk of the difference between logical and intuitive thinking. It is a very old theme: the logical versus the creative. Manipulation which is logical has the merit of precision and clarity but, by the very nature of deductive reasoning, cannot have a trace of originality. Call it structured versus unstructured, computer versus human, it is the same trade-off.

The arguments on both sides of this controversy press to the very limits – and maybe go beyond them – of what science, mathematics or philosophy can tell us on the subject. Modern thinking in science and mathematics has not settled the AI argument. Maybe science and logic will develop to the point where the Big Question – whether human thought transcends the rationality of machines – will be decided one way or the other. Or maybe the question is analogous to one of Gödel's undecidable arithmetic statements: our intuition tells us the answer but the answer will be always unprovable. Our intuition points strongly towards the notion that thinking is not purely mechanical, and that the brain is something which cannot be fully understood, let alone replicated, by applying its own mental capacities (echoes again of Gödel's Theorem). There is an old quip: 'If the brain were sufficiently simple to be understandable, we would be too stupid to understand it.' A hundred billion human neurons could in principle understand the 100,000-neuron brain of a cockroach, but the cockroach could not understand its own brain. Nor, by the same argument concerning the limitations of a process to understand itself, could we understand ours. (For the record, we do not in fact have the faintest clue as to how even a cockroach's brain does what it does.)

In a very loose sense, parallel processing and neural networks undoubtedly capture some features of the way our brains are built. But the steps which have so far been taken towards this understanding have been extremely tentative. The feature of the brain's structure which neural-network systems appear to have simulated is the principle that there are processing entities called neurons which are each connected to numbers of other neurons, which fire signals along the connecting routes (the firing even taking the form of pulses, suggestive of digitization) and, after a while, cause the connections to vary in strength. Probe any deeper and the analogy breaks down more or less comprehensively.

The first problem is in the level of complexity. Surgical operations sometimes have to be carried out on the brains of conscious patients, and through these the medical profession has found that extraordinarily detailed recollections of specific events can be made by stimulating a group of neural cells with an electric probe. An occasion such as a visit to a cinema from many years back can be recalled at a level of accuracy which would normally have been forgotten even by the end of that evening: the seat numbers on the ticket, or the clothing worn by each of the surrounding members of the audience. Where documentary collaboration has been available, these memories have been verified as accurate. Just about every event we have experienced appears to leave some sort of trace in our minds; the knowledge is ready to be applied to some future situation, even if it cannot be deliberately dredged up from the memory. With millions of visual signals coming in through the optic nerve every waking second, not to mention thousands of signals from the ears, the brain's capacity to store this cumulative information input is essentially immeasurable. According to the evidence available through microscopic examination of the neural structure, its storage capacity is indeed staggering.

As far as we can gather, the main storage points are the synapses, which appear at each connection between two neurons. Each of the twenty billion neurons is connected to the others by a tree of bifurcating wiring (the dendrites). In extreme cases, some neurons connect, via synaptic junctions, to as many as 80,000 others. All told, there are maybe 100 *trillion* synapses in the cerebral cortex. The largest processor chip has, by contrast, some ten *million* wiring connections.

More important, though, than our progress – or lack thereof – towards matching number and extent of interconnection of the neurons is the fact that we haven't the slightest reason to believe we know what goes on *inside each neuron*. The assumption that a brain-cell is comparable in processing power merely to a computer logic gate does not stand scrutiny. Each neuron is formidably complex: it contains some 100,000 different proteins, each of which itself has a highly complex molecular structure. Stuart Hameroff of the University of Arizona has been examining tiny structures called microtubules made of a protein called tubulin. Each tubulin molecule appears to be capable of more than one conformation and is thus able to represent different states (as with the switches in a computer). The idea that such proteins could be spectacularly tiny information-processing devices, each just a few atoms across, had already been offered by way of theoretical speculation by Roger Penrose. Penrose pointed out that, at this scale, the effects would work on a quantum level, meaning that they would operate not in the strict algorithmic manner of a logic machine but in the fuzzier, probabilistic mode so familiar to twentieth-century particle physics. These ideas have received further backing from Brian Josephson of Cambridge University, who won the Nobel Physics Prize for research into a quantum effect in semiconductors called Josephson Tunnelling. Josephson has since turned his attentions to psychic phenomena, which he sees as likewise being based on quantum effects in the brain.

Cells are wonders of evolutionary development. For the first three billion of the four billion years that life has existed on earth, single-celled organisms were the only living creatures. Many unicellular organisms were (and are) able to sense the environment, approach closer to nutrients, avoid potential dangers, and of course reproduce. In this light we can only speculate as to how much information-processing ability resides within the hugely complex interior of a neuron. Imagine a machine that connects twenty billion advanced information processors together with 100 trillion wires and you approach an idea of the sort of complexity we're talking about. It means that a single human being – you, for example – is in the same league as the world's total computing power.

It is commonplace to read in journalistic accounts of the digital phenomenon sentences like: 'By the year 2000 and something, the computer will be equal to the human brain in terms of memory storage.' Forecasts of this kind assume that we have got to the bottom of where and how the brain stores and processes information.

Nothing could be further from the truth. We really have no idea. Gerald Edelman writes in *Brilliant Air, Brilliant Fire: On the Matter of the Mind* (1992): 'No convincing evidence for the kinds of codes that humans use in telegraphy, computing or other forms of communication has been found in the human nervous system. The world is not a piece of tape and the brain is not a computer.' While great leaps forward have been made in neurosciences in recent years, each discovery yields new layers of unknown facts. As we focus ever more powerful microscopes onto these elements of the brain, far from converging on an answer to the mysteries of human neuron behaviour, we uncover further levels of vast complexity.

While philosophers argue about the Big Question – whether computers will ever be able to think like people – what most of us want to know is not a theoretical answer to a philosophical question but a realistic answer to a practical one:

where, in actuality, is the boundary between, on the one hand, those human mental activities which computers can effectively replicate or surpass, and, on the other, those they cannot. Could a computer program replace my lawyer – or my doctor, or my boss, or my subordinate, or any of the other people with whom I interact? Or, since I *am* a lawyer, or a doctor, or a boss, or a subordinate . . . could a computer program replace *me*?

The answer to the Big Question does not help us either way. If it is yes – there could in principle exist some combination of hardware and software which could replicate human thinking – that combination might nevertheless be so impossibly large and complex that it will never be designed and constructed. Conversely, if the answer is no – no computing device could ever even in principle pass a kind of generalized universal Turing test – we might nevertheless be able to achieve a quite workable approximation by confining the program to a topic area restricted enough – say, law – that it would nevertheless make lawyers, by and large, redundant.

Versions of the Turing test have been carried out, in a somewhat trivialized form, in which the range of topics on which the interrogator is allowed to query the computer is severely restricted. Inevitably the human questioners, although impressed by the creative programming of the machines, are able almost immediately to single out the computer programs from the humans due to their lack of 'common sense'. The computer can be made an expert on a very narrowly defined given topic – an example might be Burgundy wine – but the moment the human strays from a small set of acceptable questions the computer cannot cope.

The problem facing programmers is twofold: the sheer quantity and variety of knowledge in a human mind, and the unknown way in which it organizes and works with this knowledge. The scale of the task of encoding the everyday knowledge we take for granted is indicated by an ambitious attempt being made to tackle this, a $25 million project funded by companies including Xerox, Digital, Kodak and Apple. The project leader, computer scientist Douglas Lenat, described his goal in an interview with Michio Kaku (in *Visions*, 1998). The objective is to create a comprehensive 'encyclopedia of common sense', a list of all the propositions which we regard as self-evident and which give us the ability to get through daily life. Examples of such 'obvious' rules are:

- Nothing can be in two places at the same time.
- Dying is undesirable.
- Animals do not like pain.
- Time advances at the same rate for everyone.
- When it rains, people get wet.
- Sweet things taste good.

Lenat has so far amassed over ten million such statements, and hopes to collect a total of 100 million.

The quality we call common sense, the ability to respond to a situation in a completely unstructured manner, actually requires much greater levels of mental agility and prior experience than the highly sophisticated processing carried out by banking software or a quality-control system in a turbine-blade production line. It is hardly surprising that the expert systems programmed into computers restrict themselves to highly specific tasks.

Towards the incomparably more modest goal of simulating human expertise in suitably tightly defined areas – a particular branch of medicine, for example, or the supervision of a sales team – there have been innumerable contributions, but these have not made a noticeable dent on the need for human expertise. The facts speak for themselves. Fifty years into the computer revolution, the overwhelming majority of us tap into software that is capable of nothing more 'expert' than word processing, keeping a diary, picking up e-mail or maintaining our files in good order. If we are attorneys, our computers do not understand our legal problems for us; if we are in sales, they do not create marketing ideas for us. The whole notion that we would seek out machine intelligence to address the problems we face at work and home seems more remote now than it did in the early days of enthusiasm for artificial intelligence.

That chess game between Deep Blue and Kasparov must be put back into perspective.

The goal of beating humans at chess had fascinated many of the best minds in the computer fraternity for half a century. Claude Shannon wrote a program to solve mate-in-two problems in 1949. In 1950 Alan Turing designed a full chess-playing program; he had to run it by hand, with pen and paper, since the modern computer existed as yet only in his brilliant imagination, not as a physical object. The first working program was developed in 1956 by a group of scientists at the Los Alamos research laboratories in New Mexico, which over a decade earlier had given birth to the atom bomb and had some of the world's best computing facilities; the machine they used could execute 11,000 instructions per second. Subsequently perhaps 100,000 chess programs have been written, of which some 1,000 are of sufficient sophistication to have been matched against humans in tournament conditions.

By the early 1980s what had begun as an intellectual challenge had become big business: sales of chess-playing hardware and software were worth over $100 million per year. Technology companies began funding development on a commercial scale. IBM became a major force in the chess-programming industry and produced a machine called Deep Thought, capable of analysing two million moves a second. In 1989 it beat International Grandmaster David Levy. The final step to beating the world champion required a further eight years, many millions of dollars of IBM funding, and the construction of a new machine, Deep Blue, capable of analysing over 200 million moves a second – fifty billion moves within each three-minute period the rules permit.

A machine capable of analysing 200 million moves a second, developed by programmers with every advance in AI available to them, finally defeated a single man at what machines are best suited for: a game of logical combinations. So what? Short of the sort of ludicrously simple arithmetic task ('please add the following three numbers') that would have insulted the programmers engaged on the project, a more perfectly confined, constrained 'digital' challenge pitting machine against person could hardly have been devised.

And if, during one of the games in the New York match, the room had started filling with smoke from a raging fire, every adult and every child – even a bee with a pinprick-sized brain containing just 7,500 neurons – would have known to leave, but the computer would have gone on playing. Where in the room was the intelligence and where the dumbness? Deep Blue would have been helpless had the rules of the game been minutely changed, if its opponent had wanted to chat about the weather, if someone had asked it where to get a coffee – in fact, if absolutely anything had happened that the programmers had failed to foresee and write into its phenomenally fast but absolutely rigid logic routines. The slightest unexpected broadening of the task it faced and it would have been out of its depth. This was the machine which, in the words of newspaper commentators during that momentous week, put 'humanity to shame'.

Children playing noughts and crosses (tic-tac-toe) gradually come to realize that, if they follow a few mechanical rules, they can't lose. If the opponent follows the rules, he or she likewise can't lose – but at the same time can't win. The game becomes mechanical, requiring no strategic thinking; merely an exercise for the two players' finger muscles. Chess is like noughts and crosses; it is just vastly more complicated, so that we haven't yet discovered the rules of correct play which, when revealed, will turn it from a game of strategy into a simple exercise for the finger muscles. Chess, noughts and crosses – both are games of finite combinations, *digital* games, and digital machines will eventually work through all the combinations, or through enough of them for the rest not to matter.

A player sitting at a chessboard is thinking digital; an artist painting a landscape is thinking analog. They can both be thinking hard, and thinking well. But they are thinking very differently.

The artist is working with variables which change freely: with colours that blend into one another, with distances that can be made shorter or longer, with objects that can be freely shaped and merged into one another. Painting a sky over a rural scene, an artist may be trying to capture a certain something in the atmosphere, an ominous mood, the onset of a thunderstorm. To do so requires very careful blending of colours – experimentation with different blends of colour and shading not just of the sky but throughout the scene, altering the way the light appears to permeate the whole picture. When finally the mood has been captured, the picture almost makes the viewer *feel* the onset of a thunderstorm.

The chess player, contemplating a difficult manoeuvre, thinks in a mode known as *discrete combinatorial.* The pieces and moves are defined in discrete units. A rook is a rook, a queen is a queen, and there is nothing halfway between the rook and the queen. The player can move the rook two spaces forward or three spaces forward but may not decide, as might the artist, that it would suit the mood of the game better to move it forward two spaces *and a little further.* The analysis is called 'combinatorial' because the outcome of the game is defined by the different combinations and permutations of moves which can be made.

Real life, like painting a sky, presents us with infinite combinations. Every day we choose from a wider range of possibilities than those which have faced Kasparov during his entire lifetime of chess playing. The myriad things we can choose to do are literally not countable. That, at bottom, is the difference between the digital world and the world of Nature.

Myra Hindley is perhaps Britain's most hated living criminal. In the 1960s she committed a series of acts of brutality involving the torture and murder of small children which horrified the nation then and still lives on in its memory. The death penalty having been not long abolished, she was given a life sentence. Periodically the parole authorities recommend her release, and each time this generates a barrage of resistance in the tabloid press. The Home Secretary of the day, in whose power it is to grant her freedom, inevitably eyes the papers, mentally counts the votes and leaves the sentence to stand. Hindley is likely to remain behind bars until she dies.

In 1997 London's Royal Academy included in an exhibition a new work by Marcus Harvey which at first glance appeared to be a large drawing of the murderess's face. Closer examination revealed that it was in fact executed as a collation of thousands of tiny prints made by children's hands. There was a public outcry; mothers from all parts of England converged on the capital to demonstrate outside the gallery against the presence of this gruesome reminder of what the families of Hindley's victims had suffered a generation earlier.

No computer could scan that image and predict that

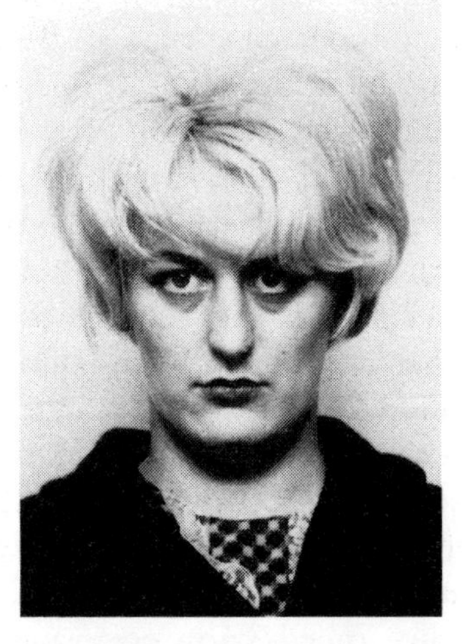

A portrait of child murderer Myra Hindley: a computer could be programmed to recognize her face, but it would never be able to understand the enormity of her crimes.

protest. There *are* visual recognition systems which would recognize that the picture depicts a human face. Perhaps a few could recognize that the drawing is made up of numerous handprints. In due course there will be software sufficiently sophisticated that it can identify these as children's handprints, not reduced depictions of adult hands. In the near future a suitably programmed machine might be able to look at the millions of faces which have been photographed by the media over the years, compare them with the portrait and identify it as of Myra Hindley; it could then look up the name on the Internet and find out what she was notorious for. But, although it could recognize each of these components of the whole, it could never understand the significance of the portrait. It could not be programmed to realize the intent of the artist. A computer could not understand what it means to be angry, nor that no emotion comes closer to being inconsolable than that engendered by the loss of a child.

Yet the public understood very well.

The processing of information into knowledge is intrinsically subjective – dependent on what the mind chooses to make of it. It relies on what is in the mind already – this is what we mean by its being subjective. Recent advances in AI technology are remarkable, but none of them has endowed computers with the richness and variety of information acquired by a human being during the course of his or her childhood. No machine possesses the prior beliefs which are the essential ingredients of human knowledge. Even if it could, an even more fundamental gap would separate the machine from the human: the different way in which the human converts the incoming information into that elusive quality called knowledge and then into that quality even further removed from digital processing, wisdom.

This is clearer now than it was twenty years ago. A computer operation is performed at the mechanical level by a machine, but it is in reality a transaction between people. Except when human programmers or users give the data meaning, a computer no more processes *information* than if its keys had been brushed by a passing dog. Information has meaning only in the context of a human interpreter: lacking an intelligent entity to interpret the information it contains, an encyclopedia is just so many squiggles on a page, no more a source of knowledge than the pattern of pebbles on a beach. How we humans have achieved our ability to find meaning in data is as much a mystery now as it was in the time of the ancients.

This fact – that information technology works at the level of symbols rather than of meaning – is the key to understanding both its great strengths and its great limitations. The chief strength of the technology is that the 1s and 0s can mean anything; the chief limitation is that, to the computer itself, they mean nothing. That the symbols can mean anything gives the machines their great versatility and flexibility; the same chips can be sold to users wishing to process information on every conceivable subject, and for each operator the device can tackle a new subject every time it is used.

The intelligence involved belongs only to the human operators and, before them, to the programmers who made the software capable of undertaking seemingly intelligent tasks by extracting only that which is analytic, logical and precise in human mental tasks. The failure to distinguish between the symbol-processing powers of a computer and the human intelligence of its operator is at the heart of many misconceptions as to the role of information technology in society.

Although the phrase 'artificial intelligence' is heard less often than it was a decade ago, it would be unfair to say, as some do, that the project has come to an end. The successes it has notched up on its way towards its ultimately overambitious goal have simply worked themselves into the mainstream.

An example of this was the development of 'expert systems'. The concept here was to isolate specific areas of expertise in people's occupations, and to simulate them in software. Traditional expert-system software had all the relevant rules built into them by the programmer, but the new AI programs were designed so that they could learn. An expert system for a bank's loan department might be fed with information about numerous past lending decisions, and gradually it would get a feel for the levels of acceptability it should set with regard to income, job stability, asset security and the like before granting credit. Expert systems are among the most significant products of the AI enterprise.

AI techniques and tools are present not just in banking

software but in engineering programs and countless other places: every time you play a computer game you are interacting with code which has benefited from ideas created in the search for a thinking machine. The writing of enormous programs like Microsoft's Windows operating systems would be impossible without chopping up the code into units and having the parts work together in ways developed in AI labs.

We will each, as we enter the new millennium, have to establish some sort of modus vivendi with these artificially intelligent creations. With technology costs coming down and labour costs going up, the economic forces point in only one direction: more substitution of people by machines. The proportion of our telephone calls which are answered by a computer, the proportion of a child's time which is spent interacting with one – these will rise as the economics of information technology works its way through corporations and educational institutions.

Whatever the fate of the 'Grand Project' of AI, its various by-products will enter our daily lives with increasing frequency. The automated directory-enquiries assistant with voice-recognition software which – more or less – catches our meaning when we ask for the number of the nearby Pizza Hut is already familiar. The slightly jerky tones in which it answers:

> The – number – you – require – is – area – code – two – one – two –

will soon become more natural as programmers master the melodies and rhythms of continuous speech; before long we will be forgetting it is a machine and thanking it for the help. The automated cinema telephone operator, which still tests our patience and our memory as it takes us through its irritating routines ('Please listen to the following list before selecting which screening you wish to attend'), will be replaced by slicker software which will understand when we just say:

> Could you please tell me if you have two seats available for this evening's showing of the new Disney cartoon, and, by the way, what time does it start?

The software, programmed to ignore the inessentials ('Could you please', 'by the way'), will pick up key words ('Tell me', 'available', 'two seats', 'this evening', 'Disney', 'start' and 'time') and give useful answers.

Then there will be more sophisticated programs: automated doctors to diagnose simple maladies, automated lawyers to give opinions, software-teachers to instruct children in geography and economics.

The current, rather rudimentary, offerings can only improve, as each generation of software builds on the previous one. Code writing has its own peculiar and highly favourable learning-curve characteristic: the financial investment, and mental effort, which has gone into producing all previous attempts make for a kind of accumulated wisdom, available, subject to copyright constraints, to the next programmer. Snippets of code, modules to perform specific functions like recognizing words, get co-opted into the next release of doctor-software. Once a problem like recognizing spoken language has been cracked – perhaps for a commercially very valuable application like directory enquiries, where telephone companies can afford to pay small fortunes for its development – the code, or at least the expertise, can be applied elsewhere.

Each incremental improvement will be welcome. If the receptionist, lawyer or teacher is to be a suite of program code, it is better that it should listen and speak to us in our natural language rather than in the multiple-choice routines of today's automated theatre-ticket vendors. Also welcome will be the fact that it is high-quality professional talent which is speaking to us through our screens. When The Learning Company or the Architectural Association put out a teaching program or a guide to home design, they will, by and large, code in the skills of good teachers and architects, not poor ones.

But, however good the teacher or architect whose skills lie behind the program design, the fundamental limitation of digital processing will cast its shadow over every such computer-based solution. This is that the teaching idea, home-design challenge, medical diagnosis or whatever else must first be reduced to the yes–no logic of the AND, OR and NOT gates which are at the core of every computer.

As programs based on neural networks, genetic algorithms and other 'nature-imitating' ideas become widespread, it will be more and more difficult to recognize the underlying strict logic of the chips which are running them. Just as in CD sound the underlying Pulse Coded samples blend into each other to render almost perfectly smooth analog tones, so the coded logic of a digital computer is almost entirely disguised by the seemingly mindlike operation of a good AI program.

Nevertheless, the intrinsically mechanical procedures are there. However creative the educators designing a geography package for 10-year-olds, every interaction between the program and the child will have to be compartmentalized by the code into the straitjacket of yeses and noes: if the child types in 'Why does the ocean have fish?' then the program can give an answer based only on its programmed routines. It can't take into account, in the way a teacher can, the child's personality, boredom threshold and other interests and come up with something unexpected, switch to a new subject or think laterally. 'Unexpected', 'new': these are terms alien to the logic of digital computing, in which everything that comes out must have been put in – by the user, or by the programmer, or by the programmer who wrote some earlier code that was here reused . . . by *someone*.

What has not been delivered is the grand finale. No machine has come anywhere near – not even within the remotest sight of – interacting with us as would a human being: nothing has passed the universal Turing test. The computer and the human brain will continue to process information each in their own, quite distinct, way.

Four billion years of the evolution of life and a few million years of the evolution of those primates called hominids have evolved a brain which has a spectacular capacity to learn, feel, drive the body and in various ways reason, but is negligibly adapted to computational or algorithmic logical operations. Some 7,000 years ago the elements of digital recording began to emerge. The complex organization of society – the introduction of money, and the use of quantitative thinking in agriculture and early engineering – made logical skills valuable, so systematic training of minds in analytic thinking became widespread; it is now the single major goal of education. Perhaps, after another million years of living in organizationally complex societies, we may develop some computer-style neural wiring to enable us to engage in computer-style reasoning ourselves. But the few thousand years during which we have lived in societies of measurement, money, and formal legal and administrative structures is not a long enough timescale for genetic evolution to have done much to our natural neural architecture. Our minds have so far remained stubbornly 'non-digital'.

6

Creating Cyberspace: Multimedia and the Internet

The extravagant mansions in Newport, Rhode Island, which the industrial barons of the previous technological era built to celebrate their wealth, are now open to the public. In the summer of 1998 I overheard two elderly couples parting after a guided tour of the buildings. 'It was nice to meet,' one of the women said, 'and we must stay in touch.' Then, quite casually, they exchanged e-mail addresses. No comments like 'Oh, so you're on the Internet!'; just the addresses: 'so-and-so@aol.com'. This brief episode signalled to me more forcibly than could a dozen market analyses that the Internet is going to be ubiquitous.

The Internet – The Net, for short – has been the information-technology phenomenon of the closing years of the twentieth century – no contest. The World Wide Web, the vast patchwork of text, images and sounds accessible through this new medium, contains contributions to meet every conceivable taste and interest. The sheer quantity and variety of information within it are bewildering: in mid-1998 it was estimated to contain about 400 million pages, but that number is growing so fast that any such statistic is almost immediately out of date.

Buddhist monks in Tibet and India – assisted by scholars in New York City and elsewhere in the United States – are embarking on the task of entering into digital format the entire Buddhist canon, consisting of 5,000 volumes containing the teachings of Buddha and the saints plus some 200,000 volumes of commentaries by Tibetan scholars. Robert Thurman of Columbia University has described the canon as '100,000 Freuds and 10,000 Jungs – a huge literature of how the mind works, how to conquer yourself instead of conquering the world'. It will all be available on the Internet. Corporations, governments, sects, political parties, clubs, and associations have each put up their stalls – their sites, to use the jargon – in this new global marketplace. They are encoding their data, their information, their knowledge – indeed, in the case of the Buddhists, their wisdom – in electronic format and posting it for the inquisitive to peruse.

Emerging as if on cue after the disappointments over the AI endeavour, the Internet swept into the new-technology arena during the 1990s. The idea of creating machines to compete with our mental powers has been quietly superseded by the development of networks for us to communicate with each other. Just a few years ago people started to use the personal computers on their desks and in their homes to access not artificial intelligence but the real thing – other people. We have come again to the idea of technology as an aid to communication – we have generated a division of labour in which people are the creators and communicators and technology is there to remove physical barriers.

With hindsight (always easy), it seems the most obvious development conceivable. There are six billion fellow human beings out there, each with more information capacity in his or her head than any in the world, and each able to interact on the broad front of human communication – thinking laterally, creatively, emotionally. Yet even as late as 1990 nobody was predicting quite how the new network would take off. Maybe we hadn't got machine intelligence quite in perspective yet – still believed that technology should elevate itself from computing to thinking and learn better to emulate our mental powers. In fact, information technology's greatest use proved to require its moving back down a step: to become a tool less for computing than for communication, so that we could learn more from each other. This is consistent with a more modest view of the role of machines and a more uplifting view of the role of people. We do not need the parallel-processing supercomputers of the AI labs, just a simple PC and a modem, to harness the creative, inquisitive and communicative spirit present in each one of us.

Well over 100 million individual computer owners have connected themselves. People are surfing this rich new source of information and entertainment, while also sending and receiving e-mails, the Net's particular mode of correspondence. Many of these individuals have sites of their own on which they post photographs, bulletins or just idle ruminations. Anything goes. Jennifer Ringley, a young American woman, has installed a camera in her one-room apartment to transmit live images twenty-four hours a day on her website; it is, unsurprisingly, one of the Web's more popular sites.

The scale and rate of uptake of this new information exchange took the computer community completely by surprise – the development provided a lesson in the difficulty of predicting what will and what will not catch on among the many applications to which digital technology can be applied. Even Bill Gates, who had brilliantly bet that the information-technology phenomenon of the 1980s would be the microcomputer and built his company Microsoft – *Micro*computer *Soft*ware – around this bet, failed to see it coming. It was only in a speech to journalists and analysts as late as December 1995 that he announced that his firm's future direction would now be geared to the Net – a reorientation compared by observers at the time to turning around a supertanker.

Today it is common to hear the Net described as the most important development in human communications since the invention of the printing press by Johannes Gutenberg in 1450. It will change radically the way commerce is conducted, as organizations replace outmoded means of internal and external communication with this powerful new electronic channel. It will change the shape of broadcasting: its websites will give a wider range of viewing than any 100-channel cable operator or 500-channel satellite reception system can offer. As for Gutenberg and his press? – well, maybe it will become unnecessary to print books any longer, since any text can now be uploaded to a site by its author and downloaded by a reader.

But what really is the Internet, and how did it come about?

The Internet was born through an alliance of two unexpected bedfellows: the academic community and the military. When the USSR launched *Sputnik* in 1957 there was concern among Americans that they were losing the technological lead in an area of potentially vital combat value. The Advanced Research Projects Agency (ARPA) was set up at the Department of Defense to ensure that the country stayed at the forefront of technology. Research and military establishments throughout the United States began acquiring computers and using them on defence-related projects. The idea emerged of connecting them together, and the Defense Department undertook the initiative to fund the construction of a data network, the ARPANET. It was this that led directly to the Internet.

For computer users to contact each other by the old-fashioned method of dialling the distant telephone number was not going to be viable, even with modems to help the bits along the wires once the contact had been established. Computers do not have conversations like people – several minutes of chat and then a goodbye. They are often online to each other for hours on end, with massive silences punctuated by a few noises as a user dabs at the keyboard. This staccato, rather chaotic, pattern of data flow needed to be reflected in a less structured approach to network use.

A new design concept was developed: *packet networking*. Computers would not dial their counterparts and wait for a circuit to be established, as we do when using the phone, but would throw bits into the network whenever they wanted in the form of small packets, each addressed to the destination computer. At each node in the network a switching device – actually another computer – would examine the packet's address and send it on to the next node. Because the links in the network are shared by so many users, the costs would then be a tiny fraction of what it would take to keep telephone circuits between each pair of users fully open.

The network is ideally suited to coping with packets destined for any point on the globe. No pair of users needs to have (and pay for) a dedicated routing for their particular dialogue, as is the case for a telephone call; the computers at each node just take in streams of intermingled packets for all users, look at each packet's address and send it on to the next node. To this day, Internet customers are typically charged a flat fee for being on the system without regard to the number of bits they throw in or how far around the world they communicate.

Back in the 1960s such a network was a gleam in the eye of several computer scientists, a gleam which was fortunately brightened by financial backing from the US Government – for a reason of little direct concern to civilians: the Defense Department liked the rather flexible mode of operation that this network of many nodes and alternative routings offered, as it would be less susceptible to enemy sabotage.

By the late 1970s, a number of private-sector networks based on the packet principle were in existence, and the idea gained ground that there should be a standardized procedure, or protocol, for them to pass information to one another. This became known as the *Internet Protocol.* In 1993 the Internet Activities Board was established to coordinate development of this new medium. In 1994 a unified global addressing system was established, and usage began to accelerate. The ARPANET was subsequently deemed to have served its purpose as the pioneering project in this technology; it was decommissioned in June 1990.

An important use to which the research community had been putting the Internet's predecessor, the ARPANET, was e-mail. Everyone who was a registered user had an area of storage (a mailbox) in a nearby computer, identified by an address of the kind gordon@caltech.edu; 'gordon' was the name he or she wished to give to the mailbox, usually his or her own name, and 'caltech.edu' was the name given to the computer at which the mailbox was located (in this case the one at the *edu*cational establishment called *Caltech*). Every time Mr or Ms Gordon logged on to that nearby computer, from a terminal at the desk or even at home, the mailbox would release any messages received.

The welcoming of non-academic and non-governmental users on to the Internet at the end of the 1980s gave the

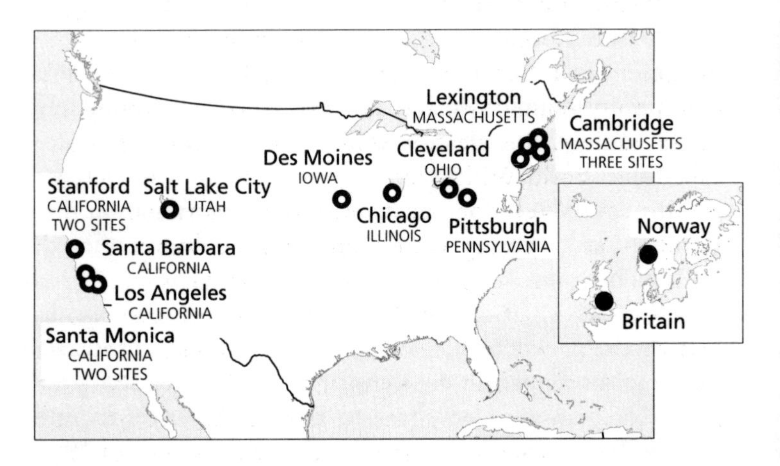

In 1971 the Arpanet, precursor of the Internet, linked just twenty-three 'host' computers in a dozen US cities. The first foreign sites, London and Oslo, joined in 1973. By 1995 there were over a million host computers on the Net.

general public access to this very useful new technique of interchanging documents. Commercial users were assigned the address '.com', government users '.gov' and other organizations '.org'. As non-US organizations came aboard they were assigned addresses reflecting the country in which they were based: '.fr' for France, '.uk' for the United Kingdom, and so on. It became possible for organizations and private individuals alike to communicate with a wide range of correspondents around the world.

It was not, however, enough to set the public's imagination on fire. Had the Internet remained an e-mail service alone it would have taken its place alongside the fax machine as a useful, cheap way to send letters and other documents electronically. The millions of businesses and individuals who are now exchanging e-mail addresses would still be exchanging fax numbers. The reason the Internet has caught the imagination of such a diverse clientele – the Religious Right, the Looney Left, schoolchildren, parents, pornographers and Senators, plus the world's entire business community – and simultaneously raised worries among most of the same is the creation within the Internet of what is called the World Wide Web.

This, the Web for short, was the brainwave of Tim Berners-Lee, a young British programmer working at the CERN high-energy-physics laboratory in Geneva. If users have mailboxes, at which they collect the private messages sent to them, why not also give them 'websites', at which they can leave public messages for anyone who 'visits' to pick up. Furthermore, there would be many interconnections called *hyperlinks* between pages of data, programmed by the creator of the site to enable visitors to jump to other websites. The hyperlinks could be to anywhere else on the Net; the Web was to be worldwide.

The visitors would be 'surfers', people going from site to site on the Internet to see what others had posted up for them to see. The Internet would cease to be just a private domain where people sent each other messages (although that would still happen): it would become a very public domain in which, in effect, anyone could become a publisher. Put up your news on your website and then just sit back and wait for visits. These would of course not be physical visits – more like peeks through a long-distance but very sharp lens. The necessary software standards were developed to enable visitors to drop into each site, take a look at a screenful of text or images, then move on to other pages of related interest.

The first implementation of the Web appeared in 1990. However, even when it was so enlivened, the Internet remained for three more years largely a province of technological enthusiasts; the commercial community and the public at large expressed relatively little interest. As of

June 1993 less than 2 per cent of Web-enabled host machines (specifically, just two computers out of a mere 130) were '.com' addresses. In that year, though, a group of students at the University of Illinois, including Marc Andreessen, co-founder of Netscape Inc., produced a browser, a piece of software which could allow ready perusal of the mushrooming number of sites and could, critically, support images as well as text. Posted on the Internet (where else?), this spread like wildfire around the world as users jumped at the opportunity to use the newly visual Web. Individuals and organizations created personal 'home pages' on which they began to publish whatever took their fancy. Initially there was no commercial objective; it was done, in the main, for the fun of it.

The proliferation which occurred from late 1993 onwards has been without precedent even in a technological field accustomed to exponential trends. By July 1996 the number of Web-enabled computers had grown more than a thousandfold, to over 150,000. The Internet became big business; by the end of 1996, 90 per cent of websites had the suffix '.com'. Today 'Internet' is a household word, and several million new users *a month* make the necessary arrangements to be on it; they sign up with an operator who can provide a gateway to this intriguing new medium.

The Internet, springing on the public so suddenly at the beginning of the 1990s, added something of a layer of confusion to the idea of the Information Superhighway. Did the Internet happen instead of it, or is the Internet the Highway? If I am on the Internet, am I on the Information (Super?)Highway?

The Net is not a network in the physical sense that the national telephone network is, with wires linking our homes and offices. It is a way of using the telephone network to get our computers to intercommunicate. It is not like the road network, a physical path to your house. It is like the postal network, which uses existing roads to deliver your mail if you make the necessary arrangements for mail to be addressed to you. The Net is a collection of computers, known as *routers* and *switches*, which are its sorting offices, and an agreed set of standards, the Internet Protocol, to which the world's computer community has agreed that information packets will conform if they are to be sent via these routers and switches.

The Information Superhighway, by contrast, is very much a physical network, an infrastructure of modern, high-speed links which will connect homes and workplaces to each other and to those computers, gradually replacing the distinctly non-super infrastructure of wire telephone links which most of us currently have running to our premises.

Being on the Internet is, for the typical user, a purely and administrative arrangement. The user contracts with a company known as an Internet Service Provider (ISP), which equips him/her with a private electronic mailbox for correspondence and a 'gateway' through which to surf the Net. The user's physical connection to the ISP – and hence the Internet – is often nothing more than his or her ordinary telephone line. Being on the Information Superhighway means having a high-speed digital connection to the world's telecommunications infrastructure. Ideally this would be a fibre-optic link, a coaxial cable or a microwave (perhaps satellite) antenna, though it could (with some compromise as to data rate) be a digitized version of the traditional copper pair. It is a real link, not an administrative one.

Somewhat by accident, and certainly with no legally constituted authority being brought to bear, the Internet has introduced into the computer field a remarkable degree of standardization. Computer installations in all sectors of industry and commerce and throughout the world have accepted that they will now intercommunicate in accordance with the protocols developed by those responsible for the Internet technology.

To get a sense of what this de facto unification has done for computer communication, it is useful to imagine what using the telephone would have been like for the past 100 years if it had been as deficient in agreed standards as the computer industry was until the Internet 'happened'. Telephones would have produced numerous different incompatible signals. To speak from your home in, say, one borough of New York to your sister in the next borough you would first have had to send a message, by mail or on foot, along the following lines:

> I am going to call you using an Olivetti telephone which produces a signal of between 12 and 24 volts, positive terminal on the black wire. Do you have a compatible one? If so please plug it in, and if not please give me the specifications of your telephone, which I seem to remember is a GTE model, and I'll see if I have something similar. And by the way, what numbering system are you on? Do you use an area code and then seven digits?

Of course, we're spared all that. The International Telecommunication Union in Geneva, through a standing body known as the Consultative Committee for International Telephony and Telegraphy (or CCITT), ensures that telephone systems throughout the world can dial each other using an agreed numbering scheme, can recognize each other's dial tones and busy tones, and transmit audio signals in the same range of not lower than 300 and not higher than 3,400 cycles per second.

Today, organizations which wish the information in their computers to be accessible to outside users do not hesitate before settling on the Internet standards as the ones to adopt.

Consequently, during the short period since the early 1990s, computers from the humble PC in a domestic bedroom to a large installation in Citibank's headquarters have found that they can, to use the telephone analogy, talk to each other.

It is precisely this ephemeral nature of the Net – that it is a feat of coordination and standardization rather than of construction – that holds the secret to both its successes and its problems. On the positive side, it accounts for the Internet's spectacular growth. To join the Internet, organizations or individuals – provided they own a computer and a connection into the telecommunications highways – do not have to build anything new. They need only install software which conforms to the standard and obtain, from the organization which controls addresses, an address for their site. When we say that there were, as of mid-1998, 100 million computers on the Net we mean that, of the much larger number of computers present in the world, 100 million had adopted the Internet protocol as the means for communicating with other Internet users.

On the negative side, the (virtual) Internet is only as good as the (real) highways it uses, and the task of completing their construction is still very much ahead of us. The links which interconnect the computers at the nodes of the network – the routers and switches – together often called the 'Internet backbone' – are of high capacity. The weak links are the ones between the nodes and the numerous small hosts and even more numerous individual users. The great majority of Internet users are not on anything remotely resembling a Superhighway right now. They are using modems to whistle and squeak, as best their modems can, their digital information along the analog infrastructure technology which earlier eras bequeathed us in the form of telephone lines.

Conversely, some people (not many) are connected to a high-bandwidth Superhighway but have not arranged to be on the Internet. Such users might wish to have a high-speed connection for, say, videoconferencing but not want to surf websites or send e-mail.

During the 1970s and 1980s, while the Internet was gradually gestating in its military–academic womb, another key trend was taking place throughout the information-technology world: digitization of the multiple analog media by which information had hitherto been recorded, stored and communicated. It was the bringing together of these two trends, networking and multimedia, which was to give the Internet the powerful potential it now has to become a general purpose 'information utility', a combination of the telephone, television and computer.

As soon as semiconductor devices had become commercially viable, telephone companies saw advantages in transmitting sounds along telephone wires as digits rather than as analog waveforms. The quality of the voice signal would not deteriorate as it passed from exchange to exchange; no longer would there be the painful and thankfully increasingly rare need to shout down a phone just because the call is going a long way.

The digitizing technique chosen was called PCM (Pulse Code Modulation). The strength of the voice signal is measured 8,000 times a second, and the measure is converted into a binary number: a string of digital 1s and 0s. Those bits encode the information of the caller's speech. There is a law of information science which states that, provided the voice signal is restricted to pitches below half the sampling rate of 8,000 times a second (i.e., to below 4,000Hz – more than three octaves above middle C, so that even the highest soprano can be heard), those bits can capture all the analog tones in the voice. The technique of converting analog sounds into digital bits is illustrated. PCM was introduced by telephone companies from the early

1970s for short links between exchanges, and subsequently spread to national and international connections.

The next challenge to digitization was video. This came much later; digital television equipment was first available commercially in the 1970s, initially only for use in broadcasting studios. The basic principle was again the same: to take an instantaneous measure of the strength of the analog signal and digitally encode the measure you get. The late 1990s are seeing the introduction of digitized television signals which reach all the way to the home. The bit speeds are millions per second – typically 100 times higher than for voice, reflecting how much higher a data rate we are equipped to take in visually than aurally; the retina of the eye contains millions of rods and cones, while the hair-like

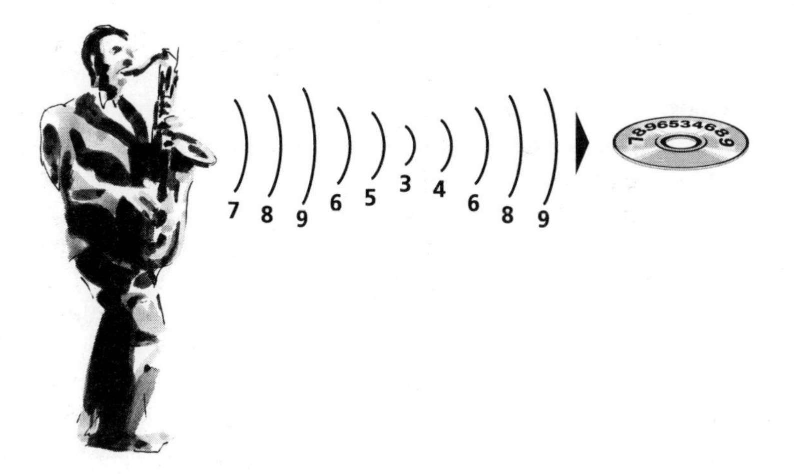

Sounds into digits: the principle underlying the conversion of natural soundwaves into bits is to take samples of the wave many times a second, and measure each sample as a digital number – ready for writing onto a digital medium such as a compact disc.

receptors in the ear number only in the thousands.

Photography is still based mainly on the traditional nineteenth-century chemical process; however, the electronic camera is now rapidly penetrating the consumer market. Of the familiar media, the one which has held out in analog form the longest is cinema, whose basic technology – exposing photographic film and projecting it onto a passive screen – has remained unchanged for 100 years. For the moment, the distribution of films is still effected by making multiple prints and physically transporting the reels of celluloid around the world. That this system remains in use, despite its cost (the copying cost alone for a typical Hollywood release is in the region of $2 million) and despite its inflexibility in the face of unexpected demand, is a testament to the difficulty even today of matching in digital form the fine resolution and colour gradation of the traditional photographic process.

However, it is difficult to envisage even the remaining analog strongholds, photography and cinema, surviving much longer as the encoding quality of the digital equivalent improves and the cost comes down – driven by the relentless law of the learning curve. The convenience and flexibility of the digital format is just too attractive. An analog photograph makes for a good souvenir to put in a holiday album, but a digitized version can be zapped to friends over the Net, duplicated at will without a trip to the photo-lab, incorporated into your next circular letter to the family and filed conveniently on your hard disk; or, if you prefer, it can be printed out and put into the album. As for feature films, when the quality level becomes adequate and the necessary encoding standards are agreed, one digitized server in a Hollywood studio will be able to send copies by satellite and fibre-optic links to thousands of cinemas.

★

The action of converting a performance of the Chicago Symphony Orchestra or a video of *Saving Private Ryan* into a stream of binary code fulfils the most basic requirement for transmission over a digital network. But it does not make it 'Internet-ready'; more needs to be done before it can be transmitted over this medium.

First and foremost, it must be divided into packets, each carrying a fragment of the orchestral performance or a moment of the movie. Every packet is prefixed by a short address label to ensure the routers along the way direct it to its destination. Packets are what the Internet is all about: if you want to benefit from the economy of this medium (to Paris or around the corner, all for the price of a local call), from its flexibility (send anything by e-mail – words, sounds, pictures), you cannot expect the network to set up a separate channel for transmitting your particular message. You must put it into standardized packets and throw them in with everyone else's.

The second element of preparation, an optional one, is to compress the stream of bits so as to reduce its demands on channel capacity. A human voice encoded by the PCM technique requires 56 kilobits per second to carry it in uncompressed form, and high-quality CD music is even more voracious – the rapidity of sampling of the music signal is as high as 44,000 times a second, and the precision of the coding requires 16 bits per sample, so that hundreds of thousands of bits are generated per second. This is much faster than the data rate most people can achieve when accessing the Internet via their telephone connections.

What comes to the rescue here is the phenomenon of *redundancy*: the fact that the visual and audio signals which Nature produces contain a great deal of repetitive data which is redundant for the purpose of conveying the sights and sounds we see and hear. To recreate the sound of the horn in an orchestra requires a very high bit rate if we insist on separate samplings of the sound every millisecond. However, the player is not in reality changing the sound that fast, so the successive samples contain a great deal of repetitive information. Software of great sophistication has now been written which takes advantage of the similarities in successive samples of a digitized sound recording to reduce the number of bits required to encode each minute of an audio signal.

In 1995 a relatively little-known company based in Seattle, Progressive Networks, launched its RealAudio System. This allowed music to be compressed and played over the Internet at just 8 kilobits per second. The sound quality was tolerable, certainly to Web enthusiasts, who were delighted to be able to listen to music 'live' via their modems. Rapidly, newer RealAudio System releases and competitor products appeared that improved the sound quality; at the same time modems got faster, so the bit rate which could be accepted by an average user rose to its current level of over 56 kilobits per second. Compression does still result in some loss of quality – the compression tricks will never quite catch the transition between two horn notes with the precision of the original signal – but the results are generally perfectly acceptable.

A much bigger problem confronts the transmission of video material, which has a prodigious appetite for channel bandwidth; uncompressed video in broadcast studios runs not at thousands but at millions of bits per second (over 140 million as it leaves the cameras). The target of the compressing endeavour is to squeeze an entire television programme or movie along a pair of copper wires such as connect most homes to the telephone system. This would result in an automatic upgrading of the telephone network, and hence the Internet, to full video capability.

This challenge, something of a Holy Grail for telecommunications engineers, is being addressed from two directions. The first is by way of enabling the copper wires to carry more bits – as many as possible – so creating out of the humble analog telephone link to your house and mine something known as a Digital Subscriber Line (or DSL). The current view is that somewhere in the region of a million bits per second could be carried along many of the world's copper pairs without too great a proportion of the binary digits being lost along the way – a massive improvement on the rate at which bits are spurted out even by today's best consumer modems.

The other direction of attack is to compress the video signal, certainly down to the level which could be transmitted by a Digital Subscriber Line. Ideally it should be compressed much further, to the bit rate of a modem, creating universal access to video material over the Internet without any further improvements to the telephone infrastructure: television pictures and films would be available at the end of any phone line.

Although on the face of it compressing a digital video image from many millions to just, say, 50,000 bits per second presents formidable difficulties, they may not prove insurmountable. There is a very great deal of data redundancy in a video image; a typical television image contains large areas of, for example, blue sky or background room colours, none of which need be painstakingly reproduced, millimetre by millimetre, across the width and height of the image. Similarly there is a large element of repetition from frame to frame as the picture is rescanned many times a second. Provided the *true* information content of a video image, as defined by the entropy measure, does not exceed 50,000 bits per second, the signal could in theory be compressed into such a channel. The engineer who discovers how to achieve this ratio of compression while maintaining a broadcastable picture quality will have discovered that Grail. It has not happened yet – video so compressed is suitable only for projecting inferior-quality images on the small screen of a videophone. Nevertheless, the rudiments of video-by-telephone are with us and, with time, the quality can only improve.

As a consequence of the digitization and compression of voice and video, the Internet is becoming a multimedia channel. Once a soundtrack or moving image has become a stream of bits, it can be posted on a site on the Net and 'visited' by any other user. So the distinctions between the media by which we communicate with each other are becoming blurred. Why talk only on a telephone, watch broadcasts only on a TV, or work at your desk only with a computer? Particularly as Internet applications compete for 'eyeball time' with both broadcasting and computer pastimes, manufacturers of both computers and televisions are trying to persuade customers that theirs is a general-purpose device which can be used for the whole gamut of screen-based activities.

NEC sells a product called WebSync which listens to the soundtrack of a television set while the viewer is watching and automatically looks up related material on the Internet. With the appropriate hardware, software and service subscriptions, the consumer can request that synchronized pages of related information from the Web appear alongside the TV picture. Microsoft and a number of large electronics manufacturers have also entered this market, the basis of which is to couple intelligent processing to the information sources available on the Web to give an Internet-enhanced television service which can be tailored to the preferences of each viewer. Over time, its software analyses the user's viewing preferences and accordingly adjusts the balance of programming presented. It also searches the Internet for relevant facts and figures which can enhance the quantity of information available on the subject being broadcast –

additional statistics on a player's performance during a sports match, for example.

The logical endpoint of multimedia is virtual reality. When the senses of not just sound and vision but touch and feel are transported, so is, in principle, the 'sense of being somewhere'.

With the development of the technology of analog communication based on the forces of electromagnetic propagation uncovered in the nineteenth century, and of the technology of digital communication made possible by semiconductors in the twentieth, there is now no reason in principle why *any* sensual impression should not be created in one location, digitized, transported to another, and offered to a recipient. The sensual impression could be that of a real thing, like a video camera panning a horizon, or an entirely synthetic one, like an artist's impression of the scenery on a nonexistent planet. Take any item of information – a news reporter's camera shot (sight), a musician's chord (sound), the pressure detected by a robot arm (touch), the output of a gas spectroscope (smell), the formula for saccharin (taste) – encode it into digits, and you have it in perfectly standardized format: stimuli for each of the five main human senses, and much more if you wish, converted into strings of 0s and 1s. The conversion itself can be a major technical challenge, as the complexity of a video camera attests, but, once it has been done, the bits can be stored in any digital medium, transmitted over any digital network, and processed in any digital computer.

Computer scientist Jaron Lanier doubles as a virtual-reality performance artist. The musical instruments he plays as he contorts around the stage do not exist. He wears wrap-around goggles and a data-glove, and the Sax, Xylo and Rhythm Gimbal are virtual images, cyber-instruments produced by a computer and projected into his field of vision. As his fingers grope at thin air, tiny sensors in the glove know what imaginary keys he is touching in the world he sees through his goggles, and the processor backstage fills the auditorium with electronic sound. The audience sees, projected on a giant screen behind him, the same virtual realm that he does.

The audiences at these shows are only *half* immersed in the virtual world of cyber-instruments which the computer is creating: they also see a real performer, on a real stage made of real wood. The next step, when goggles come down in price (which they will do in the next few years – virtual-reality kit is still on the costly side, except for very low-resolution entertainment gadgets), will be for each of the viewers to wear a pair too. Then they will be as immersed as the performer, and maybe could be allowed to join in. Equipped with data-gloves, they could press a few virtual Xylo keys, taking care not to bump the real-world fan in the real-world next seat.

'Virtual reality.' The label is the perfect oxymoron, for the opposite of 'virtual' is 'real', so what are those of us without the goggles seeing? Real reality? But the term makes its point – we can block our eyes and ears from seeing and hearing what is really there, and substitute something which is not.

The ideal, which science can conjure up but which engineering has yet to perfect, is complete sensory immersion: not just a data-glove for the hand, but a whole body suit equipped with thousands of pressure pads, with a nose pouch into which a computer-controlled chemistry set delivers scents, and so on. With every sense organ covered, the body can pretend to be anywhere, seeing and touching and feeling anything the programmers dream up.

In the final phase the audience can stay at home. Since what will go into and out of their goggles, gloves and other paraphernalia is nothing other than digital data, why not network it? You and I will be able to take part in a virtual-reality performance broadcast from anywhere in the world. Slip on the headset, dial up and join in – welcome to cyberspace.

A programmer in Switzerland, a group of students in Illinois, technology buffs in Silicon Valley and Massachusetts and Seattle, driven by messianic zeal, adrenaline and caffeine, all with hardly a thought for commerce (though many have become hugely wealthy), launched an idea which is now causing leaders in the world's communications industry to rack their brains as to where it is leading. By virtue of the Internet and its World Wide Web, a new modality of communication has emerged.

Before, telecommunications provided us with circuits and channels – telephone circuits and broadcasting channels. When on the Net we do not dial up circuits or tune into channels. We enter a space. This is a new kind of interconnectedness. We throw packets into that space, and fish packets out. One packet for a keystroke or mouse-click, a thousand for a minute of telephone chat, a million for an hour of video. We will have to get used to the idea of this space, in which information is not present in a specific location that you must physically connect to, like the office of American Airlines, or present at a specific time that you might have missed, like the six o'clock news, but is just there, accessible wherever and whenever its owner wishes. When William Gibson adopted the term *cyberspace* he had in mind this realm in which information 'floats', only loosely tied to its many originators.

A dimension has been added to communications which neither Gutenberg's press nor Marconi's radio transmissions could provide. The Net has removed the asymmetry, the yawning divide, between producer and consumer, and this is indeed a completely unprecedented feature in any mass medium. On the World Wide Web each of us can be a publisher and a broadcaster as well as a reader and a viewer. All Internet subscribers who take the trouble to put a page up on their website, whether a political manifesto or just a smiling passport photograph to let the world know what they look like, are publishers with a global reach of millions of readers.

The creation of the Internet has taken networking technology out of the corporate bunker and put its power into the hands of people in all walks of life. Access to information has become available more broadly than at any time since the advent of mass literacy. Anyone who can get access, perhaps just for a few hours at a time, to a computer and modem – which means almost anyone in the developed world who has a modicum of initiative and motivation – can find out as much as they could possibly wish about almost literally any topic.

Economic and legal implications of such a change in information accessibility are beginning to emerge. Information has a particular property known as *non-depletion*: if one individual passes it on to another then, unlike the case with physical goods, the quantity of it possessed by the first individual is not depleted. This property is both very desirable – any number of people can enjoy a novel – and at the same time problematic, in that theft of intellectual property is difficult to trace since nothing disappears as a consequence of the theft.

As information technology advances there are periodic scares that it will make impossible the enforcement of copyright, or indeed any control over the unauthorized dissemination of information. The photocopying machine was feared by book publishers, the cassette recorder by the music industry, and video recorders by the movie industry. Those crises faded away: authorship rights were (and still are) abused, but the problem is a manageable one. Now there is a new threat. Once an item of information has entered this 'space' called the Internet, control over it is lost. A piece of intellectual property can reverberate around the world in seconds and proliferate exponentially. Each owner of a mailbox can act as a redistributor of whatever he or she finds there. If a sufficient number of 'Netizens' choose to make something public knowledge, it will within hours be reposted to host computers around the world, accessible by anyone without payment, without fear of censorship, and without respect for privacy.

The new medium is (largely) frontier-free and censor-free. Because it is a way of using the existing telecommunications network, not a network in its own right, there is no way to control what is published or broadcast – or read or viewed – by a country's citizens. The content is just a stream of bits, indistinguishable to the frontier guard or censor from the billions of bits of non-Internet traffic travelling as radio waves or along wires into and out of that country. The wires could, in principle, be cut – albeit at the expense of cutting off the country from the rest of the world's telecommunications networks – but the radio waves couldn't be. As computer guru Esther Dyson wrote in *Release 2.0* (1997), 'everyone will be on the Net, except for a few holdouts'. The Net will be everywhere there is a telephone line, or a cellular network, or a path to and from an orbiting satellite; for all practical purposes, it will be everywhere, period.

The distinguished political scientist Professor Ithiel de Sola Pool predicted early on how electronic technology was moving in a direction which would serve the causes of liberty. His last book *Technologies of Freedom* (1983), tracked the history of telecommunications and pointed to the way in which the proliferation of technical means of reaching around the world, coupled with the convergence of media brought about by digitization and new network architectures, would make censorship all but impossible.

With analog electronic media clearly destined for the scrap-heap of history – the vinyl record is gone, the non-digital cellular phone is being replaced, digital television and photography are on the rise, and now the strictly digital Internet is sweeping all before it – it is worth recapping what digitization does and does not achieve. Exactly why do we have this preference for converting the natural sounds and images of our human sensory channels into 0s and 1s?

First, digitization does not make fidelity or quality *intrinsically* better. In fact, strictly speaking, the initial consequence of digital coding is to make a sound or picture worse. It changes the smoothly changing waves of sound and colour which enter the microphone or lens to inevitably 'jerky' digital code. The technical terms for the (two) sources of problems are *sampling error* (the distortion arising from the fact that a finite number of samples is taken from an original which is changing continuously) and *quantization error* (the distortion introduced because each sample is coded as a whole number or quantum, rather than as an infinitely flexible measure to reflect the infinitely variable intensity – quantum-mechanical quibbles aside – of sound and light waves). In other words, an analog recording of a particular phrase sung by Luciano Pavarotti can, at least in principle, catch all the nuance; a digital encoding of that phrase is imperfect firstly because it measures only samples in time, not every moment in time, and secondly because the samples are not measured exactly right.

Richard Solomon, an MIT-based researcher, has been engaged on a project to develop an extremely high-resolution electronic camera – of a quality that could, for example, reproduce Old Master paintings to near-perfection. Needing to transmit the signal to collaborators in California, he ordered an *analog* coast-to-coast line from telephone company AT&T. This is seriously high-fidelity work: no converting to bits please, only pure analog transmission will do! And there are clubs of audiophiles who claim to hear sampling and quantization distortion; they eschew CDs in favour of carefully maintained vinyl records and reel-to-reel tapes. I have tried listening to Pavarotti both ways and can make no similar claims: for me, and I suspect for most listeners and viewers, digitization does not detract from the fidelity of the analog original. But it certainly does not add to it, notwithstanding publicity to the contrary by cell-telephone or TV-channel vendors.

Secondly, digitization does not set aside the laws of physics which constrain the capacity of a radio wave, wire or fibre to carry information. Maxwell's laws operate with the same rigour on a stream of bits as on an analog wave. As compression techniques improve, the amount of information we can pump down a telephone wire or television channel will increase; technology has already advanced to the point where, typically, four digital television stations can fit into the airwaves previously allocated for one analog station. Four, yes, but forty, no. The constraints are still there – and, unfortunately, the greater the compression the worse the digitization-induced distortions.

Where digital coding wins out is not by producing an improvement in the quality of the sound or picture transmission but in preventing further degradation of the information as it passes through imperfect media. The sampling and quantization errors are one-off distortions introduced at the moment of digital encoding; from that point onwards there is no reason a single bit should be misplaced until it reaches the destination – the phone handset in Indonesia, the television set on a ranch in Idaho, or the CD player in an apartment in Moscow. Digital storage and transmission do not produce gradual degradation, like the fading colours on celluloid of such concern now to movie archivists, or the progressive deterioration of a long-distance phone call as it made its way across a continent in the 1960s. A bit is a bit and, provided the signal is not so swamped by interference as to change a 1 all the way to a 0, or vice versa, the information will maintain its integrity.

Then, most intriguingly, there is the prospect of digital processing – not just processing the bits as a stream of uninterpreted *data*, as in compressing a video for enhancing speed of transmission, or chopping it up into packets so that it is Internet-ready. We refer here to 'real' processing, processing of the *information content*, of what the bits represent. Do we not hear how computer enhancement brings clarity to blurred pictures coming from Mars, or helps make sharper the face of a rapist caught on the security cameras of a shopping mall? Is not *processing the content* where digital really scores over analog?

As with transmission, the answer is no in principle but yes in practice. In principle, enhancing a picture or identifying what it contains can be done on either the analog original or the digitally encoded version. Photographers have traditionally touched up their pictures of fashion models to *take out* the lines under the eyes; an instrument could be programmed to *put lines onto* a criminal's picture to make the facial outline clearer. Programs have been developed to guard the young against exposure to indecent pictures on the Internet by recognizing large expanses of the colour of naked flesh. An analog instrument to pick up colour tones would be equally effective (or, actually, ineffective: this technique is easy to circumvent – change the colour when transmitting, change it back when receiving). In practice the inexorable progress of chips – Moore's law increasing their complexity and the learning curve driving down their costs – guarantees that digital processing will be the way forward.

But we should not expect too much too soon. Even when

information is encoded digitally, the fact remains that, as far as the user is concerned, the two big objectives which information technology set out to fulfil earlier this century remain separate. One (communications) is to deliver data from one point to another as faithfully – that is, as little changed – as possible. The other (computing) is engaged with precisely the contrary task: to change one set of data into another. So sweeping has been its success in harnessing the power of computing that it is easy to forget that the Net is nevertheless quintessentially a medium of *communication*. Although the use of the Internet involves a combination of computing and telecommunications technology, the computer processing is there only for the purpose of aiding in communication; whatever happens at one end – by way of compression, adding packet addresses, etc. – subsequently 'unhappens' at the other: the compressed signal is uncompressed and the packet addresses are stripped out.

The gap between the power of digital technology as a tool for communicating information and as a means of making sense of this information is tellingly revealed when one trawls the Web for information to tackle a specific need. A by-product of the standardization of the Internet is the possibility of searching it automatically for relevant data. A number of organizations, such as the highly successful Silicon Valley start-up Yahoo!, offer websites which serve as computer-compiled catalogues of as many pages of different sites as they can identify. These catalogues are becoming increasingly sophisticated in their ability to respond to information queries; though lacking the mental agility or plain common sense of a human counterpart, they contain 'search engines' which scour the Web for keywords which might give clues to the content of a site, and have built up enormous databases of what is 'out there'. The problem faced by users who type into their query-boxes that they want to find out about a selected topic – be it anything from the Ku Klux Klan to Paris fashion – is not generally that the search engine draws a blank but that it comes up with hundreds or indeed thousands of sites of little relevance.

Digitization is a characteristic of the way facts are encoded – not of what they mean, of their importance, of their relevance or of their value. We must always keep in mind that this is fundamentally a technology of transmission, an efficient means of accessing knowledge which must originate ultimately with people. The Internet is a new way of accessing *content*. The content itself is what people and organizations choose to put on their sites – the 5,000 volumes of Buddhist teachings, the marketing presentations of a business corporation, or the personalized contents of the websites of millions of users.

The quality of human creative output is not primarily dependent on the quality of the available technical facilities; the content is greatly more important than the conduit, and here digital technology has much less to offer. Here the supply of creative talent, not the technology, is the bottleneck.

A movie – or a television channel or a book – is technologically a flow of data which can, for convenience, be digitized. In human terms it is an interaction between the director – or producer, or editor – and the audience; at this level the digital tools don't offer much. In spring 1995 the Sony Corporation wrote off $50 million when it cancelled its unsuccessful experiment with interactive cinemas. The idea had been that feedback from the audience to the screen could be used to allow movies to have different endings, depending on viewer preferences. The technology for making the end of a movie change depending on viewer preferences had become available and so, naturally, it was tried out. Whether it was through poor market research or just through a failure of imagination, the Sony executives had ignored a fact known to the tellers of Icelandic sagas a millennium ago: the very basis of engaging the hearts and minds of an audience is that it is the storyteller, not the audience, who decides how the story will end.

Audiences do not adapt to improved technical media if the content is not to their liking. There is a long-running tradition of failure in high-tech 'interactive television'. Early in the 1990s Time–Warner ran well-publicized interactive cable-television trials in Orlando, Florida. These proved so unable to stimulate the required interest in the audience that shares took a downturn throughout the nation's cable industry.

Video (i.e., visual) software began to be created a hundred years ago and fifty or more years back had already been developed to the artistic and creative standards we enjoy today; some modern movies can equal *Gone with the Wind*, released in 1939, in their impact, but surely they do not surpass it. The real investment in making the sort of programme or movie that billions of viewers will watch is in the rare creative talent of the scriptwriter and director, and in the vast fees commanded by the stars.

The enormous quantity of information available through such computerized networks all has to be created and processed into the right format, and then it must be interpreted and understood. These tasks have become neither easier nor more difficult as a result of the Net; any attempt to have them performed by computers is still subject to all the practical and conceptual constraints which have thwarted the attempts to replicate by machine the human thinking process: the lack of creativity, of common sense, and of intuitive reasoning.

Nowhere have these constraints been more clearly signalled than in what is now known as the productivity paradox: the puzzling failure of information technology to transform the level of economic output. To this we now turn.

7

Looking for Results: Computers and Economic Progress

In the elegant St James's area of London is a shop which has not changed in nearly two centuries. It makes shoes for those who have a sense of tradition, a desire for the best, and above all money – a pair costs over £1,000. Hand-made wooden lasts, carved to the exact shape of each foot, are kept for the distinguished clientele – Baroness Thatcher, leaders of industry, and many aspiring to such ranks. On a wall is a framed letter from Buckingham Palace in which the Duke of Edinburgh thanks the shop for sending sixtieth birthday wishes. 'Perhaps I am keeping so well,' it says, 'because I am so well shod!'

From the outside the store appears to be just like others on the street, but step inside and you witness a scene of eighteenth-century manufacturing – every step in the process of turning cow-hides into finished products, awaiting customer collection from a traditional glass-fronted mahogany cabinet. When a client enters, the proprietor finds the carved wooden last of his or her foot (it is never discarded during the customer's lifetime) and selects a cow-hide from the stack on a bench in one corner of the workshop. Over the next six weeks cow-hide will be turned into a new pair of shoes. Each step in the manufacture is done by hand, as it was before the days of factory production. Then a bill – an 'account', they still say – is written out, in ink, for settlement at the end of the quarter-year.

If all production were still performed by hand like this, firms would have little use for office information technology. What need would there be for computers to write out bills or for databases to maintain customer lists? The 'office' aspects of the owner's work – to write out a customer's account after six weeks and deliver it by the next quarter – is trivial by comparison with the 'real' task of making the shoe. The proprietor's equivalent two centuries ago would have laughed at the idea that he could have a machine to write out bills; he would have far preferred one that produced shoes.

The world had become very different during the two centuries since all shops were like this. The first half of the twentieth century saw huge improvements in the economy's ability to produce and distribute manufactured goods. Between 1900 and 1960 the United States experienced more than a fivefold increase in the output of each industrial worker. The traditional goal of economic activity – to satisfy the population's needs for housing, clothing, food and durables – was being achieved with an unprecedented degree of success. The Industrial Revolution had increased immeasurably the efficiency with which every activity in the economy was undertaken.

Except one: information handling. The economy was ready for a machine which could process information – the computer.

Investment in the new technology has grown accordingly.

The expenditure by US businesses on office technology has overtaken that on all the traditional capital equipment used for the *physical* processes of production and distribution – in factories, in the transportation industry, in the production of chemicals, energy and power, on farms. There is a great deal of data on this; the US Government's *Survey of Current Business* shows the ratio of information to production equipment spending was about 1:5 in 1970. Then investment in the new technologies began to spiral upwards. By the end of the 1980s an important symbolic threshold had been crossed in the United States: the amount of technology investment for white-collar personnel, at over $100 billion, was now the same as that for production staff. The most successful industry of the late twentieth century was now firmly entrenched in the homes, and above all in the offices, of the industrialized world.

Today, there is nowhere quite like Silicon Valley: 25-year-old men (occasionally women) find themselves courted by Wall Street money in a place where, the saying goes, shake a tree and venture capitalists fall out. A cheque is signed, the new gizmo (or new code) is launched and a year later the entrepreneur is a flourishing corporate executive. The United States is riding a high-tech-driven boom centred on dynamic hardware and software industries which support thousands of enterprises developing and deploying user applications.

Since the invention of the chip in that seminal year of 1958, the industry of information processing has developed beyond the wildest expectations of its early inventors. Equity markets put exceptionally high valuations on companies which are seen to be set to take advantage of the demands of an information-hungry society: the two Stanford University engineering students who founded the Web directory company Yahoo! received financial backing of $1 million in

1995; within three years the firm was valued at $2.4 *billion.* Internet-service provider America Online has a market capitalization of $25 billion against a shareholder capital value in the accounts (that is, book value) of only $125 *million.* The mysterious realm of bits in silicon appears able to produce miracles on the stock market to match any in the laboratory.

However, as technology expenditures race ahead it has become clear that improvements in the performances of computers at the data-processing level are not being matched by comparable improvements in the performance of an increasingly knowledge-based economy. As of 1990, all was not well with the performance of the new machinery. Throughout the first several decades of the twentieth century economic output per head of population in the industrialized world had grown rapidly, achieving a fourfold increase in the period up to 1950 as factories and farms were transformed by technology investment. From 1950 to 1970 average productivity in the world's five largest industrial economies accelerated further, reaching the unprecedented annual level of 5 per cent. Then, just as computers entered the workplace in significant numbers, it slowed down, to languish at about 2 per cent for most of the next twenty-five years. In the United States, the unquestioned leader in information-technology investments, annual productivity growth for the twenty-year period to 1995 averaged only 0.8 per cent.

The question was naturally asked: why do the very large expenditures on computer technology not have an effect on economic output in the massive and obvious way that the purchase of production technology did earlier in the century? This productivity paradox, as it came to be called, generated a great deal of controversy and soul-searching among scholars, business executives and government bodies. Many excuses were given. Some wrote of measurement difficulty: the benefits of the new technology were mainly to be seen, we were told, in service sectors like finance where productivity is difficult to quantify. Some commentators asked us to wait a little longer: the computers and networks have been bedding in, and the really big benefits are just around the corner.

Not very convincing. Twenty-odd years of computer spending before the worthwhile returns set in is not the way companies like to run themselves, especially when announcing their quarterly returns on Wall Street.

A cloud was looming over the otherwise remarkable story of the computer industry. While the new technologies were undoubtedly driving a boom in the industries which *produced* them, and produced *for* them, there was no evidence of a decisive, demonstrable impact on the performance of the industries which *used* them.

Although analysts have gone to great lengths to discover decisive evidence of a link between technology investment and productivity, the goal has eluded them; and there has been much support for the contrary position. Paul Strassmann, for many years head of information at Xerox Corporation, concludes in his *The Squandered Computer* (1997) that 'there is no correlation whatever between expenditures for information technologies and any known measure of profitability'. Professor Michael Scott-Morton of MIT finds that 'evidence at the aggregate level does not indicate any improvements in productivity or profitability' due to information technology. Stephen Roach, chief economist at Morgan Stanley, specializes in the information-technology sector and has studied and surveyed this field for many years. His summary is unambiguous:

> The productivity gains of the information age are just a myth. There is not a shred of evidence to show that people are putting out more because of investments in technology.

Robert Solow of MIT, Nobel Laureate economist, is equally blunt in his assessment: 'We see computers everywhere except in the productivity statistics.' What is most telling about this productivity paradox is that there is any controversy at all – that after some thirty years of such enormous outlays (much larger now than investment in the technology of industrial production) it is still necessary to scrutinize computer investments so carefully to find even partial confirmation of their overall economic impact. The effect of mechanization on the production of domestic appliances, fabrics, the chemical industry or cars was utterly self-evident, and was measured in *multiples* of increased output; in the case of office technology we are struggling to confirm *percentage* improvements. It seems that information machines have had much less impact on economic output per dollar or yen of investment than would be the norm for investments of this magnitude in the traditional area of industrial production. To understand why, it is necessary to look at the way data, information and knowledge are produced and consumed in a modern economy.

That information is of great importance in human organizations is clearly not new. The phrase '*Scientia potestas est*' (Knowledge is power) was coined by the English philosopher Francis Bacon in 1597. The fortune of the Rothschilds, Europe's greatest banking dynasty, was built on access to timely information; history's most famous carrier pigeon is the one this family sent to London with news of the British victory at the Battle of Waterloo on 18 June 1815, yielding the owners massive trading gains on the stock exchange. What is new is not that we use information but that we use it in such prodigious variety and quantity.

Ages of history can be defined by the predominant activity which consumes economic resources. At the end of the eighteenth century the great majority – 87 per cent –

of the US workforce was engaged on the land growing food; this was the final phase of the agricultural period in human history, a phase that had lasted for several thousand years, being preceded only by the hunter-gatherer era. During the course of the nineteenth century the industrial era became established: employment in industry grew from just 2 per cent of the total workforce in 1800 to over 30 per cent in 1850, and by 1900 industry had outgrown agriculture as a source of income and wealth. Finally came what we now know as the post-industrial age.

The business pundit and scholar Peter Drucker stated as early as 1968 that the transformation in the way in which information is managed in the economy constitutes a revolution that may have economic consequences as large as those brought about by the Industrial Revolution. Kenneth Arrow, presenting the lecture of the winner of the 1973 Nobel Prize for Economics, spoke of information as 'an economically interesting category of goods which have not hitherto been afforded much attention'. This was beginning to change. During that decade Fritz Machlup of New York University was working on a ten-volume study entitled *The Production and Distribution of Knowledge in the United States*; and in 1977 a very detailed analysis of the production and distribution and use of information, *The Information Economy: Definition and Measurement* by Marc Porat, was published by the US Department of Commerce; it concluded that 51 per cent of the American GNP was devoted to strictly information tasks.

A picture of a modern information society emerged. The economic system can be said to have two components: a production activity – growing, building, manufacturing and distributing physical goods and services using the limited labour and material resources available – and an information-processing activity, which involves managing, organizing, coordinating and developing the many individual productive activities. The information-handling tasks include clerical work, accounting, brokerage, advertising, banking, education, research and other professional services. Their production counterparts include construction, mining, manufacturing and agricultural work, and service activities such as transportation and the operation of restaurants and hair salons.

Occupational data give useful measures of the proportion of the economic resources of the US economy that is devoted to overcoming the information problem. A person falls into the 'information-processing' category if his or her output has value because of its information content – as in memoranda, decisions, financial transactions or research reports. All those personnel concerned with the creation, manipulation, distribution or servicing of tangible physical goods – factory operatives, drivers of vehicles, construction workers, craftsmen or waiters in bars – fall outside this category, being treated as production workers. The proportion of the working population concerned with information handling has been rising dramatically, from some 18 per cent at the turn of the century to more than 50 per cent now. The message comes across even more strikingly if expressed in terms of the ratio of information to production work: in 1900 production workers outnumbered information workers almost five to one, but by 1980 the information workforce exceeded the workforce which was sustaining the production base of the society.

It is difficult to accept that more than half of all our human resources should be going into an activity as abstract and ephemeral as the manipulation of knowledge – that it should be engrossed in the virtual world of information rather than that of physical objects. The 'goods' being produced – items of data emanating from offices, computers, databases – are not only abstract in the physical sense, having no tangible weight or substance, but are also often of only fleeting usefulness. So much of the information economy seems to be engaged in producing documentation that will be read just once, instructions that will be irrelevant once executed, accounts that add nothing to the value of the products to which they refer. Data does not have the comforting solidity of a piece of furniture, the durability of a house or the direct utility of an item of food or clothing. Information is always an ancillary – it is always *about* something.

But the figures are unambiguous. In the 1770s production tasks had begun to be automated in the English Midlands, the cradle of the Industrial Revolution. Two centuries later, businesses were spending more on information about things than on producing the things themselves. The days in which shoes were made by hand and the office overhead was a minor additional consumer of resources were over.

It was at this point that computers were able to start working their silent magic on the huge quantities of numerical information which had begun to engulf the organizations of the developed world. During the 1960s large companies and financial institutions rapidly adopted computers to automate many clerical functions, as did governmental bodies. The early transistorized computers were perfectly suited to the task. Though they were expensive, they could perform thousands of operations per second and of course could calculate and sift data with complete precision.

One of the unglamorous consequences of the explosive economic growth of the early and mid-twentieth century had been the need for large numbers of clerical staff to perform routine mechanical tasks. Big corporations and public administrations contained offices stretching to seeming infinity in which sat row upon row of human data-crunchers, putting marks onto payroll slips, stock-control forms, bank-account statements and address labels.

The repetitive mental act required was typically no more than to transfer a number or word from one form to another, occasionally performing a simple arithmetic operation on a mechanical adder that sat on the desk. The unchanging routines of these white-collar production lines were as mind-numbing as those on the factory floor, if not more so; the physical environment was probably better, but there was not even a glimpse of the shoe or vacuum cleaner to which the blue-collar operative was making some tiny but tangible contribution.

In the 1970s and 1980s the inroads of technology into the white-collar workforce continued – indeed, accelerated. But an important change began to take place in the types of application being addressed. Whereas initially computers had been used to undertake essentially computational tasks, they were now applied to the less structured activities which go on in a typical office environment. A new challenge loomed: once the data-crunchers – those small armies of clerks filling out forms by hand in the insurance offices of the 1960s – had been replaced by computers, it turned out that the great majority of white-collar personnel worked at what we label the levels of *information* and *knowledge*.

The Black–Scholes Formula is run in the computers of traders dealing in financial options markets. It calculates the price they should pay for an option on a stock. For example, Du Pont stock has reached $20 and you have a hunch it will rise further. You don't want to tie up money in it now, but you are willing to gamble a small sum to have the choice to buy it at a later date. How much should you lay out for that option – $2 per share, maybe $3, maybe $4? Before the formula was published in 1973 options pricing was intuitive. Myron Scholes of Stanford won the 1997 Nobel Prize for Economics for bringing computation to this problem (his collaborator had died two years earlier). He and Fischer Black had set about the task of calculating how money works in an options market in just the same way that a chemical engineer might calculate how fluids flow in a petrochemical plant – by using mathematics. The options industry has been transformed; the Chicago Board Options Exchange makes over a million trades a day, up from a thousand on the day it opened in April 1973. This is an example of the notion that knowledge can be *engineered*. In time, it is hoped, a great deal of the information processed in business would be amenable to such treatment.

Yet the computer was to have surprising difficulties in pursuing its assault on white-collar productivity.

In a research study by John Gould, a human-factors expert at IBM, office professionals from within that company's research organization were given eight letters to compose. Four were to be written with the help of word processors, and four using pen and paper with a secretary then typing the fair copy. The purpose was to see whether the use of computers either increased the quality of written output (as assessed by judges) or reduced the overall cost. The results were negative on both counts. Cognitive scientist Professor Thomas Lindauer found this result to be the norm when reviewing the many available studies of office technology productivity for his book *The Trouble with Computers* (1995). The text of a letter or business memorandum is readily digitized. However, that does not make the letter or the memorandum the *product of a digital operation*; it is the product of a human mind, and it is *there* that the cost and quality are primarily determined. There is a yawning gap between data shuffling, on the one hand, and thinking, on the other.

The study confirmed something that many of us feel instinctively as we sit at our desks: that, although the new tools are undoubtedly desirable (who of us would now give them up?), they have not revolutionized our output in the way that machinery revolutionized the manual production of pins, as famously described by Adam Smith in the eighteenth century. A business letter is a new piece of knowledge. The task of committing it to paper is one of data manipulation, and here the computer can help. But the computer's contribution is relatively minor by comparison with the mental task of deciding what the letter will say.

It is easy to be misled into thinking that a tool which produces data is also producing knowledge. The alphabetic characters and numbers produced by an office printer describe the ideas contained in a memorandum, but they are not the ideas themselves. A computer does not produce reports in the way a production line manufactures pins. A pin-manufacturing machine automates the production of pins, and that is the end of the story. If it automated only the production of the packaging in which the pins were to be delivered to the customer, leaving it to a craftsman to produce by hand the pins themselves, then its productivity impact would be like that word processors have had on report-writing – useful, but not revolutionary.

The distinctions between knowledge and information and between information and data can be carried through from the micro-world of symbols and signals to the macro-realm of human organizations. The easiest but still the best way to get a grip on the pattern of information use in business is to look at what people do. Occupational statistics – statistics on what people do for a living – are very detailed. An occupational description – medical secretary, advertising executive, retail store manager – provides a good indication of how each member of the workforce adds his or her small bit to the hugely complex pattern of information interaction within the economy. This quantitative estimation is made possible by the very high degree of occupational specialization in modern societies. The US Bureau of Labor Statistics classifies the workforce into over 400 occupational categories. Using these statistics it is possible not only to distinguish between information and non-information work but to get some feel for the type of information-processing work which non-production workers do.

Such occupational data confirm that only a small minority of 'information occupations' deals with data in the routine and mechanical way that a check-out clerk handles money at a supermarket till. The overwhelming majority is engaged in the administration, coordination, monitoring and organization of economic life in ways which vary from task to task and from day to day. Even less routine is the work of people in research and development institutions, in education and training establishments, and in creative professions such as writing. They are building up the long-term stock of knowledge available to the next generation.

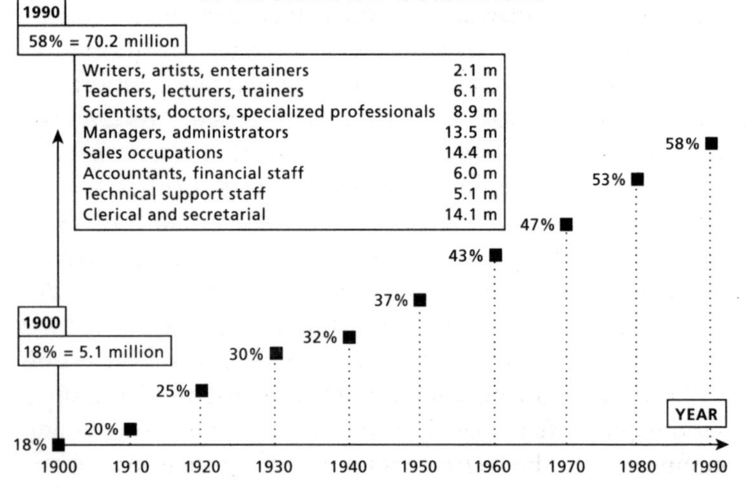

The information revolution has been first and foremost people-driven, an explosion throughout this century in the proportion of the population that works with knowledge and ideas.

The modest impact of new technology on productivity reflects the fact that not all information work is equally susceptible to automation. A legal opinion is generated by the mind of a lawyer; this underlying product is not susceptible to automation by information technology even though the technology will cope well with the processing and delivery of it once it has been encoded as alphabetic text. Banking may seem to be all about numbers but at bottom it is still about borrowing money from customers and finding safe assets to invest it in.

In short, the great majority of 'information workers' do not spend their days doing what could by any stretch be called computing. As the computer community was to discover, the next level of human information handling – processing – was going to be a lot more difficult to digitize than communication had been. And thereafter the highest level, thinking, was going to prove impossible.

There has always been a balance struck in business between quantitative and humanistic skills. Harvard Business School and MIT's Sloan School of Management inhabit the same smallish city of Cambridge, Massachusetts, and there is a famous culture clash. MIT has many lectures in techniques, like finance and management-information systems. The Harvard course is based on cases – discussing business situations, emphasizing human skills, presentation, interaction. It is joked that, when students bring their piled-up baskets to the shortest check-out lines in the local supermarket, the cashiers point to the Five Items Or Less sign and ask: 'Are you at Harvard and can't count, or at MIT and can't read?'

Another quip runs among the professors, and it makes a serious point. It is that the MIT students come out ideally equipped with the techniques they need for their first jobs in middle management; the question is how they will cope when they reach the board fifteen years later. By contrast, the Harvard MBAs are ideally suited to the jobs at the top;

the problem is whether they will manage to muddle through the intervening years.

Most critical decisions, in business as elsewhere, are not reducible to computation. The engineering of knowledge, as in the Black–Scholes Formula, has turned out to have only a partial application.

In finance houses throughout the world, analysts study screens showing the prices of stocks, bonds, currencies and commodities; computers work out trends and arbitrage possibilities and the analyst makes a recommendation to a client. A good decision can earn the client $100,000 and the analyst a welcome bonus. On 'Black Wednesday', in October 1992, the billionaire New York financier George Soros was assessing if the British Treasury would get enough support from the French and German authorities to withstand a run which had developed on the pound. He talked with three colleagues whose wisdom he respected, decided it would not, and bet massively against sterling in the currency markets. The pound was forced to devalue and he made a profit of more than $1 billion in one day. He had used no technology aside from his telephone.

Few industries have more use for office technology than do financial services – banking, insurance, brokerage. Lloyds of London, in the spectacular stainless-steel headquarters built for it by the arch-modern architect Richard Rogers, had all the computer equipment one would expect of an organization about to enter the twenty-first century. It was brought low in the late 1980s because it had exposed itself to the wrong sorts of risk, especially US claims relating to asbestosis. The problem was the old-fashioned one that has been at the heart of the insurance business since the sixteenth century – guessing right about what can go wrong.

Yamaichi Securities, the brokerage firm that in November 1997 became Japan's biggest corporate bankruptcy, was likewise at the cutting edge of technology. As with many other

Japanese financial institutions, it exposed itself to the wrong sorts of risk and then engaged in the unwise – and illegal – practice of covering up trading losses. The systems tracked the trades with great sophistication, but were powerless to know what was really going on.

Computer technology has improved immeasurably the quality of information management in the business world: the sheer quantity of data had certainly been heading out of control and would by now have been extremely onerous without the new devices to keep track of, and sort, it in millions of offices. But it has not caused a wholesale displacement of people from the workforce – people who on the surface may have seemed just manipulators of data, ripe for being made redundant by the new technology, but on closer examination turned out to have had more to contribute as creative or understanding human beings than as crunchers of numerals and letters.

Most of what goes on in the business environment does not involve entering financial or stock-keeping data into a ledger, nor working out mathematically how performance is comparing against growth or percentile targets, but is much less structured in format. Above all, a huge proportion of it concerns dealing with the organizational interaction of people. As soon as a digital information-processing system encounters the vagaries of the human mind, the software and hardware requirements to handle any interaction become so overwhelmingly complicated that they will either require an entirely new, and not currently foreseeable, generation of information-processing capability or, according to some schools of thought described in this book, will be intrinsically unable to cope.

The growth of the Internet has re-energized the idea that a technology-based transformation of the economy and of society as a whole is imminent. Lou Gerstner, chairman of IBM, made the following statement in his 1997 commencement speech at Wake Forest University:

> We're now riding the next great technology wave: the rise of powerful global networks like the Internet. Something very important is happening here. Networks are collapsing the physical barriers between nations, markets, cultures and people. This connectivity will change everything: the way we access entertainment and interact with one another.

Hopes are high for a network-led solution to the problem of sluggish productivity growth. The technology industry has introduced the idea that we are entering an era of 'e-business'. This encompasses a wide range of ideas, from the increased use of electronically supplied information for taking business decisions to the integration into a single automated system of the cycle from customer enquiry to order fulfilment. The talk is of a comprehensive restructuring of the distribution of goods and services around an electronic infrastructure. Those companies which provide customers with the most seamless route from website visit to confirmed order – tailored precisely to the desires of the prospective user – will win the prizes of the e-business era.

Worldwide, for reasons in some cases related to information technology and in others unrelated, markets are more competitive now than they were a generation ago. The stock-holder model of corporate ownership is replacing family and state ownership, putting greater pressure on financial performance. Many more managers, equipped with business degrees and ambitious to climb the corporate ladders, are willing to conquer new product and geographic markets. On the buying side, consumers, informed by more media channels and increasingly international in orientation, are willing and able to choose from a greater variety of suppliers.

In this new period of exceptional competitive pressure, a key weapon in the corporate arsenal has been technology, and more especially information technology. The reaction of business managers to this latest round of technological development has been to invest more heavily.

The amount spent on information-technology support for the US economy in 1997 was an extraordinary $225 billion. The ratio of expenditure by US businesses on technical support for the office as opposed to industrial processes, which, we recall, had reached 1:1 by the end of the 1980s, has now shot up to 2:1 in favour of the white-collar sector. The automobile factories of Detroit are spending more on information technology to support the office functions of general management, marketing, customer communications and accounting than on the much-vaunted automation of the plants themselves. US hospitals invest more on computer technology for administration than in the wards and operating theatres. Other advanced economies either show a similar imbalance or, lacking quite such a large investment in computerization of offices, are hurrying to catch up.

Professor Eric Brynjolfsson of MIT, who specializes in researching the productivity impact of IT investment, in reviewing the techniques used by major companies to justify such expenditure observes that 'they reveal surprisingly little formal analysis'. Corporate executives seem to rely on the belief that there is something intrinsically different about technology investment, which couldn't – and therefore needn't – be subject to the usual rigours of investment justification.

The 1990s continued to witness curious phenomenon: computer investments rising at record rates while scholars and pundits hoped for but could not find any formal justification for this. It seemed that in many otherwise prudent corporations budgetary discipline disappeared whenever any project included the word 'computer' in its title. The entire automation enterprise was also, of course, helped along through its catching a favourable point in the business cycle – the 1990s were boom years for most of the developed world.

Insofar as time is money, and electronic linkages can help eliminate the delays and errors which are inherent in an organization as complex as a modern economy, electronic networking quite definitely does bring benefits to corporations and their customers. The Automotive Network eXchange (ANX) is a network-based communication system installed in 1988 to link the Chrysler Corporation, Ford Motor Company, General Motors Corporation and their top 1,300 suppliers. The Industrial Technological Institute, an automotive-research organization, has estimated that the reduction achieved by ANX in delays in passing information along the supply chain about car seats alone is worth about $71 per car to the industry – a total of a billion dollars a year.

The airline industry derives particular value from good linkage with customers because the product – an airline seat – cannot wait for the next flight if it is not earning revenue. Typically a large modern airline runs ten different fares on a plane; with 3,000 flights a day serving 30,000 pairs of cities, that makes for many millions of combinations of routes, prices and load factors to optimize. American Airlines, which pioneered the computerization of this task in the mid-1970s, values the resulting benefits at an extra $500 million per year in revenue. United Airlines has now installed a supercomputer based on the technology of IBM's chess-playing machine Deep Blue to handle its reservations: probability analysis, applied statistics and econometrics are used to analyse months of past bookings, cancellations and fare records to determine how to plan the discounting and advance selling of the 160 million seats on United Airlines' 850,000 flights of the upcoming year.

Furthermore, it is now impossible to maintain the respect of customers, suppliers and most especially staff without being demonstrably up-to-date on computing. Every 'information worker' must be equipped with a personal computer, digital mobile device and Internet access. Whatever the Morgan Stanley productivity analysis may say, the perception is that to have the latest tools on the desktop is vital; and, to a degree, perceptions are reality. For that reason alone corporations would be ill advised to reduce technology investment – and are not about to do so. Corporations cannot escape the need to keep up with the latest in computer and networking technology any more in the last quarter of the twentieth century than they could escape the need to keep up with industrial technology in the first three-quarters.

However, when assessing the potential for electronic networks to transform business efficiency it is important to keep in mind that such technology has already been available for over twenty-five years, and that where businesses have needed it they have applied it. The banking sector started to enjoy the benefits of a worldwide packet-switched network in the 1970s; this network, SWIFT, automated a large proportion of inter-bank transactions. Proprietary versions of the same networking concept were applied within banks for their own use, and similarly within large corporations such as General Motors. The world's airline industry, again since the 1970s, has had a network called SITA for reservations, ticketing and general inter-airline correspondence. By the end of that decade no large corporation was without a computer data network linking its various facilities; many had started to link in selected suppliers and customers as well. Professional groups with special needs for organized information sources were also wired up to computer networks by the mid-1970s: a database called LEXIS served the legal community, MEDLINE linked physicians, and a variety of commercial services such as Compuserve (which twenty years later would be a major Internet service provider) offered various business-information services. Reuters has been serving the financial sector with up-to-the-minute stock and currency data for a quarter of a century.

What the Internet has done, very dramatically, is to provide businesses with the same quality of linkage to the market at large as they used to have within their private corporate or industry networks. But the business world has not just been waiting around for the Internet to happen. It has financed the creation of a powerful set of media for getting through to customers, and has developed hugely sophisticated techniques of getting to the hearts and minds of those customers – and hence to their wallets. These techniques – advertising, product positioning, brand building – will be no less applicable in the networked marketplace. The Internet provides a new infrastructure for market contestants to fight in, but does not change fundamental rules of the game.

The problem in obtaining a customer is not principally one of having technical access – by telephone, post, Internet or otherwise – but one of winning the customer over. Here matters like branding, loyalty, persuasion and of course the value of the product on offer hold sway. Likewise, the problem in managing subordinates is not mainly that of maintaining technical contact, by e-mail, cellular phone and voice-mail, but that of offering understanding, motivation and guidance. In the post-Internet world there will still be much more to managing and motivating subordinates than improving their access to information, and there will be much more to entering a new market than posting your offering on the Internet and hoping for custom to arrive.

Employees must be persuaded to give of their best, and customers to part with their money: neither is a task for a computer.

The physical means of access must of course be there, and if technology improves those means that can only be for the good. What we must not forget is that the most important means of technical access among players in the economy, the telephone for voice and various technologies for text (telegraph, then telex, then fax), are quite mature, and had their effect on economic structure and performance decades ago. When information workers have needed access to information, they have had much better access to it, and shown much more creativity in getting at it, than today's enthusiast of e-business may realize. As the modern form of business organization has built up during the twentieth century it has developed intricate webs of information flow to satisfy the specific needs of individuals at their workplace.

A major airport at the close of the twentieth century is a monument to human organizational achievement. At peak times Chicago's O'Hare International Airport takes in a plane every ninety seconds; aircraft wait in giant circular stacks which form above the city before being guided into their place in the precisely defined line of landing craft, approaching the runway three degrees off the horizontal at eight-kilometre intervals. When the schedules are tight, each incoming airliner, representing tens of millions of dollars' worth of aeronautical engineering, has to be brought in, led to the gate, refuelled, checked over, serviced and guided back for takeoff within forty-five minutes. Even more problematically, each plane takes off bearing on average 150 passengers. These passengers need to arrive at the airport, present their tickets, check in their luggage, receive their seat allocations, obtain refreshments, do any last-minute shopping, pass security and get to the right gate.

Despite all this, the overwhelming majority of flights depart on schedule.

Not surprisingly, the airline industry has a voracious appetite for information technology. So have the automobile industry, the manufacturing sector, the extractive industries, energy and chemicals – and above all the information-products sectors like banking, insurance and commerce.

A technically advanced economy is an organization of millions of independent decision-takers, consumers and producers transacting with one another in a staggeringly complex web of relationships. Information-flows are the lifeblood of an organization; in a sense they are its very essence. It is a sign of how mammoth is the task of managing these flows and relationships that a modern economy has half of all its employees working in purely information-handling environments. In such an economy it is not surprising that telecommunications and computers are seen to have a transformative potential.

But, although it is often stated that the defining element of a modern organization is the flow of information, the focal points of economic and social organizations are in fact *people*. Even in this information age organizations are still based, primarily, on interactions between people: employers and employees at work, parents and children at home, buyers and sellers in the marketplace. Machines, however powerful their data-processing capabilities, have not taken over as anchor points of economic, governmental or social organizations, and they are not able to interact with the human anchor points. They do not function autonomously within human organizations; they are not able to exhibit that key attribute of the member of a human organization: the ability to 'deal with people'. That is why computers based on Turing's logic-machine principle, however fast they become, will remain only partly successful in their ability to substitute our skills.

We must not overestimate the speed with which an organism that has the huge complexity of the global economy will reorganize its patterns of interaction in response to information technology. There are two changes taking place: a rapid change and a slow one. The rapid change is in how each individual person is using technology to go about their information-processing work, as falling prices make new applications like personal computing and Internetworking viable. The slow change is in how the organizational structure of the economy itself – as measured by the numbers of individuals engaged in each category of knowledge-related work – is transforming.

The president of, say, the Exxon Corporation has an almost completely non-routine set of problems on his mind for the simple reason that, with such a pressure of responsibilities and with such resources at his disposal, anything even remotely routine or instruction-based is delegated far further down the organizational pyramid. What is left to concentrate on is a combination of strategic thinking and, even more important, the task of holding together an organization comprising a highly diverse group of people. It is this which gives the corporate leader's job its variety and imposes its mental demands. The same goes for political and other organizational leadership. There are many offices in which we currently find people sitting where at some time in the future we can envisage opening the door and being confronted by a computer, but the suite of the chief executive of Unilever and the office of the Governor of New York are not among them.

What is true at this rarefied managerial level is also true, albeit to a lesser extent, further down. A check-out clerk in a supermarket registers the price of each article in the shopping trolley, causes the machine to produce a total, takes money from the customer and returns change. This activity falls almost entirely into the 'blind' data-handling category (today's tills even work out the change to be given, obviating the only major demand on mental arithmetic which the job of a check-out clerk used to make), so we can confidently predict the rapid erosion of this occupation by technology. But we should always be careful to recognize the non-routine or non-algorithmically predetermined activities which even a check-out clerk will occasionally do: getting up to help an elderly shopper who is having difficulty handling a package, reassuring a customer as to the freshness of the oranges or the validity of the sell-by date on the ice cream, and so on. There will always be a human attendant in the supermarket, even when the vast majority of today's check-out workforce has disappeared.

While technology is improving spectacularly in speed and capacity, the most fundamental bottlenecks in information production and consumption are in the minds of the knowledge producers and consumers. Microelectronics technology, following Moore's law, is halving the cost of data-processing power every two years; at this level – the technical – the economics of cyberspace is following previously unseen exponential cost–performance trends. However, at the level of *content* – of the knowledge economy – there are unchanging fundamentals.

The Labour Theory of Value is a concept in classical economic thinking which originated in 1662 in a treatise written by the English economist Sir William Petty. He observed that the main determinant of the exchange value of a commodity is the amount of labour which goes into its production. This concept, much refined over the centuries, can still serve as our guide to where value will lie in the information age. The computer is commoditizing the processing of digital data, not of human knowledge, and if it is not bringing the expected gains it is because what goes down in cost also goes down in value. The value has stayed, so far, with the creative energies of people. There is more subtlety in these than the technologists and economists initially credited.

On closer scrutiny, then, the productivity paradox begins to be resolved. There have been two information revolutions. One is in the processing and transmission of data by machines; the other, much more gradual but ultimately more important, is in the production and use of knowledge by people.

The failure of information technology to produce a decisive uplift in economic productivity or in the quality of consumer entertainment has taught us a striking lesson which it would be well to bear in mind during the next round of technology deployment. It is confirmation, by way of a worldwide, trillion-dollar trial, that it is and will be much more difficult to automate what we do with our minds than it was to automate what we do with our hands. We should see this not as a failure of technology but as a tribute to human skills.

8

Back to the Real World: Digital Technologies of Tomorrow

The brief history of information technology has reached the turn of the millennium. Digital electronics has become, unquestionably, the leading technology of this era, and it is inconceivable that it will do other than develop massively in the twenty-first century. It is a good time to take stock, and ask what has been achieved so far, before speculating as to what the future might hold.

In the transmission and processing of sounds, images and symbols, the technology industry has made progress beyond the dreams of its pioneers back in the 1950s. On *transmission* of the raw data of the human senses, the application of digital techniques to the previously analog world of sounds and images has been an almost unqualified triumph; technology scores high marks, close to ten out of ten. The human body has the capacity to take in about 20kHz bandwidth of sound data through the ears, and some megahertz of bandwidth of visual data through the eyes. Taking into account the fidelity of the ear and the information-carrying capacity of the optic nerve, an information channel of several megabits per second comfortably exhausts the ability of the human observer to take in (let alone process) information physically. A studio-quality audio system reproduces music and speech across the 20–20,000 Hertz range of human hearing with almost perfect linearity (evenness); a casual listener can easily mistake the reproduction for the real thing. Although video is still short of that level of quality (a glance at even the best professional monitor is enough to reveal that one is watching an image and not reality; a combination of flicker, resolution and lack of three-dimensional feel give the game away), the simulation is good enough to be thoroughly satisfying.

In North America, Japan and Western Europe most homes, and a large proportion of offices, are already linked by coaxial or fibre-optic cables. At present most of these cables are not equipped with either switching or two-way transmission electronics, thus limiting their use to one-way broadcasting applications, but soon such premises will be able to have two-way facilities which can deliver data to individuals in their homes at rates vastly in excess of those which the biological sense organs – including the eye – can absorb.

Already, the owner of a digital cellular phone has a near-ubiquitous communications device: in most parts of the developed world where a person is likely to be undertaking a professional engagement, he or she can transmit digital bits representing voice, data or graphic information to practically any inhabited place on earth. The only commonplace application for which the capacity would at present be inadequate is video. With the development of compression technology and the opening up of more of the radio spectrum this constraint, too, will disappear.

On *processing* speed and capacity, technology again scores

close to ten out of ten for the purpose of most individuals' present requirements for digital manipulation of information. In the 1990s, as microprocessor clock speeds reached and exceeded some 100 million bits per second, PC users became aware that for the first time the reaction of their machines to the majority of commands was essentially instantaneous. When engaged in word processing, diary or address-book management – and provided that external access to networks or slow-memory devices is not involved – the response to user instructions occurs typically within a fraction of a second; that is, as fast as a human interlocutor can, physiologically, have need for.

The computational resource available at the fingertips of any home or office user with a moderately advanced personal computer – say, from the Intel Pentium chip upwards – can undertake logical processing operations such as calculation, manipulation of words and diagrams to a level which massively outstrips the performance of the human brain. Indeed, the same could be said of much more primitive chips, going all the way back to the Intel 8080. Admittedly there is still much more that could be done to speed up and enhance the handling of full graphics, and to move faster through the vastly complicated software packages (with over ten million lines of code) which are being provided to even relatively undemanding office and home-computer owners – as producers strive to keep at the forefront of the industry by satisfying a very wide range of possible user requirements. However, at the rate at which processing power is advancing – still at least doubling in speed and memory capacity every two or three years – it will be several years at most before even the cheapest computer has processing power which enables it to respond within a fraction of a second to the full range of commands which the average user is able to master, remember and key in.

Just as in the case of communications, where the capacity

problem is effectively solved when the technology can accommodate any input or output speed which the human sender or receiver can biologically handle, so the need for further improvements in processing performance will be constrained not by the speed of the chips but by the capacity of the human mind, individually or in organizations, to interact with the machinery. Again as with communication, we may be within sight of an important milestone, namely the point at which as much processing power is on every office desk and in every home as the user can reasonably need.

It is not possible to quantify just how high a processing speed a PC needs before it is no longer offering increased value to its user, but intuition suggests that we are approaching that limit. Those who doubt this should remember that in this field what is in today's professional catalogues is in tomorrow's homes. The Cray T3E-9000 is the first commercial supercomputer to break the barrier of one trillion flops (FLoating point Operations) per second; in the time it takes you to blink, about a fiftieth of a second, the Cray will manage thirty billion calculations. Professor Nicholas Negroponte, head of MIT's Media Laboratory, has pointed out that the advertisements for computers which our children will read will not describe their processing power – it will be taken for granted that this will be adequate. He draws an analogy with the modern automobile market, in which television advertising emphasizes increasingly marginal accessories rather than horsepower; the ability of a car to travel at any speed which the driver could reasonably handle is self-evident.

The latest in this series of advances is digitization: the conversion into the computer language of bits of the formerly analog media – telephone, radio, television – and also of information which was not previously in electronic form at all – printed matter, personal letters, doodles, whatever.

Information in homes and businesses which was previously in analog format – from the handwritten memo to the photograph-based slide show – is being encoded digitally, making it capable of storage, transmission and processing on the premises or, over networks, worldwide. Once the signals have been converted to digital, we can experience them, like a Jaron Lanier virtual-reality performance, anywhere. The Internet is the most important embodiment of this trend; its standardized protocols are replacing the diverse approaches to digital communications which existed in the past. The Internet is becoming the framework within which the digitization of telecommunications will be completed.

Looking back over the past 150 years, it is apparent that progress occurs not at an even pace but in big steps. After each key new technique is developed, there is a long period during which the population adopts it and adjusts to its presence.

After the Morse telegraph, the big new step forward was the telephone, introduced in the late nineteenth century. From its early days its function was clear: to allow conversation over a distance. There is a completeness, a finality, about that function, which is, not surprisingly, still the telephone's basic and defining purpose. In 1981 a public lecture at the Massachusetts Institute of Technology was given by scientists from the world-famous Bell Laboratories, the foremost research organization in the field of telecommunications. A member of the audience rose to lament the fact that the telephone had changed so little in fifty years; the speaker responded, to laughter from the auditorium, that another thing that has not changed in fifty years is the distance from the ear to the mouth, nor the tone and intensity of our voices.

Another big step was the development of radio and then television broadcasting. During the 1950s and 1960s, the proliferation of television revolutionized the access of people to 'lifelike' information on developments around the world. The TV was instrumental in introducing a new pattern of domestic life, whereby people typically began to spend hours each day in front of the set. As with the telephone, the domestic television had a quality of completeness; those households which acquired one fell into the habit of daily viewing and, although the quantity of channels has changed greatly, that pattern of use has remained broadly unchanged since. Once the time spent in front of the screen by the average Northern Hemisphere household reached some four to five hours daily (figures vary according to definitions and methodologies used), which it did more or less as soon as television reached satisfactory technical levels in the 1960s and 1970s, it did not move up further. Viewers who now have access to hundreds of channels are watching no more television than they were when there were just two or three.

Like the provision of a telephone to carry the human voice or of a television to bring moving images into the home, the whole digitization enterprise also has an end – a point of completeness. Once the process of conversion to digital is done, it is done: you can't make a sound, image or document 'more digital'.

And thereafter? The mix of essentially limitless communication bandwidth and processing capability will make for a powerful cocktail. In the Institute of Medical Psychology in Tübingen, Germany, a software package analyses the 'brainwaves' picked up from an electro-encephalograph (EEG) wired to electronic sensors placed on the head of an experimental subject. Hundreds of volunteers – almost all that try – are able after several hours to move the image of a rocket on a computer screen in any direction they choose, simply by thinking. The translation from thought to movement is still primitive – for example, the software can be programmed to shift the rocket to one side if the volunteer thinks of something sad, while another emotion will make it move up or down – but the interpretation of brain signals by computer has begun.

Microsoft is providing $80 million funding for a laboratory next to Cambridge University in England. This will not be a laboratory with smoking funnels and whirring machines, but will consist of rows of desks at which programmers sit writing software. The hardware innovations will be concerned with the link between the machines and our human senses. Nathan Myhrvold, the firm's top technologist, explained that the work there will seek to develop computers which have what is termed a *natural interface* with the user – in his words, computers which can 'see you, hear you, understand you'. Natural-interface technology is consuming most of the company's $2 billion a year research program; the aim is to have linguistics abilities by the year 2002, and also that the machines should be able to recognize gestures and facial expressions. IBM has similar objectives; a spokesman talks of 'a world where using your computer is as easy as talking with your best friend'. The company has demonstrated a computer which can be controlled by a combination of speech, gestures and movements. (Not thoughts – yet.)

Countless changes will confront us as we learn to take advantage of this new networked world.

It is safe to assume that within a small number of years – less than ten, perhaps – we will have the technical ability to retrieve instantly, from anywhere in the world, essentially any item of information whose owner has chosen to make it accessible. You will be able to say into a mobile phone:

> Contact five travel agents in India who can book a beach holiday in Goa, and send them all a message asking them to give me quotes for a two-week stay in a three-star

> hotel. Also, when I get to my hotel room tonight I want to see *Les Enfants du Paradis* with English subtitles. Oh, and check my refrigerator at home for beer and, if there's none left, alert my car navigation system to give me the route to a supermarket which is open late tomorrow.

The computerized service to which we are speaking will put the various instructions in hand and, very likely, will succeed in fulfilling them. We will gradually – and sometimes not so gradually – come to realize how much of everyday life was taken up with tasks which comprise quite rudimentary information processing. In the retail world of the future there is no reason why you should not just pick a shirt off a clothes-rack in a store and walk out with it, a chip in the shirt communicating to the store that the item is being bought and a chip in your watch instructing your bank to pay the store $20 for it. For purposes of security, a further chip in your watch might be able to ensure that it was you who were wearing it, not a watch-thief. Examples like these are commonplace in the computer literature, and rightly do not generate controversy. The ability of digital machines to undertake manipulation of data of this kind in almost unlimited quantities is simply no longer in question. They will put at our disposal an ever richer virtual world, an increasing ability to manipulate and process symbols into images and sounds which will inform and entertain us.

Yet the nagging fact remains that we inhabit not a virtual world, of code in networks, but a real one, a world of chemistry and biology, of industry and Nature. Whether in addressing economic performance, tackling ecological renewal or even just providing entertainment, the cyberspace of code in networks can do only so much. There are limits to the ability of images on screens or facts in databases to cater to our needs. The world they are presenting is not, in the final analysis, the real one, and we will always feel that difference.

The encoding is far from perfect: what passes through networks is only part of sensing. In addition to the images which can fit through the 'window' of a computer's screen and the sounds which can be reproduced on its speakers, we are sensitive to an extraordinary variety of background sights, noises, odours and tactile sensations which make the reality surrounding us so much richer than the finest multimedia simulation.

For example, the human nose, even though it is an extremely poorly performing organ by the standards of most mammals, plays a critical role in our overall perception of a location and our sense of well-being in it (we remember many places by their smell, even from childhood; and people who have lost their sense of smell – a rare medical condition – report a consequent loss of interest in food and sex). Its more than 1,000 receptors, operating in combination, allow us to distinguish some 10,000 smells. In principle, it is possible for odours to be digitally encoded – a product called Neotronics Olfactory Sensing Equipment (NOSE) has been developed by a collaboration between the University of Warwick and commercial interests in the UK – but the complexity of the chemical factory which we would need in each of our living rooms to reproduce scents for our delectation defies the imagination.

More problematic still will be the full simulation of tactile sensations. Engineers have developed data-gloves with motorized pressure-inducers which can give a limited tactile feedback, but the ultimate goal of a kind of mechanized body stocking which simulates the pressures and pains felt by the hundreds of thousands of receptors on our skin is currently way beyond reach.

We may not consciously formulate the thought but we display clearly through our behaviour that we are keenly aware of the qualitative and emotional distinctions between a real experience and a virtual one. Take entertainment.

Warner Bros. have recently bought a derelict power station in the heart of London and the bulldozers have already moved in – not to demolish the huge structure but to convert it into a complex of cafés, restaurants and riverside walks – plus thirty-six cinemas. The investment is worth hundreds of millions of dollars. Across the Atlantic there is talk of a new renaissance in Harlem, reminiscent of the renaissance of the 1920s. The streets are cleaner, the shops are picking up, and crime is down. By night, new jazz bands are playing and people – not just from Harlem, but from midtown Manhattan – are packing in to hear them.

If the predictions made in the 1980s had been correct, Warner would not be spending its money investing in city-centre real estate, and there would not be an influx of custom to Harlem bars: the fibre networks coming to our homes would be showing us all we wanted to see, and in the comfort of our own lounges. Cities were more or less written off as the smog-filled detritus of a pre-electronic age. So far, though, despite the extensive diffusion of the multimedia technologies of CD-ROM, computer games and cable–satellite television, people are resisting the temptation to immerse themselves in these media at the expense of getting out and about. More time is being spent at the computer terminal, yes, but this is at the expense of television viewing hours, which have actually fallen during the 1990s, rather than 'real life'. Meanwhile, evenings out have started to pick up; the 1997 survey *Cultural Trends*, published by the UK Policy Studies Institute, shows that, although over 80 per cent of the population have at least one video recorder in the home, people are going to the cinema, theatre and opera, and visiting museums, *in greater numbers than in the 1980s*. The same trends are recorded throughout Western Europe and in the major cities of North America.

The people filling the jazz bars in Harlem and sampling London's evening life are sending a simple message to the technology industry: there is only so much we wish to take of the view through the screen; we also want to partake of the real world – warts, crowding and all. We have a deep attraction for the real, and mediated communication, through videophones and the like, is just not the same. We may tire of scrutinizing images for yet more hours on computer terminals, or of looking at the renderings of faces of actual or often imaginary people. Our sensual antennae are much too complex to be satisfied by the electronically generated simulations reaching our eyes and ears from a desktop workstation. There is vastly more going on in our interaction with the world than any electronic delivery system can simulate.

A radical alleviation of urban overcrowding – and particularly the end of the problem of congestion caused by commuting – is another of the revolutionary societal implications of information technology to have been widely predicted. Equipped with satellite dishes linked to a broadband Internet, we can live in high-tech latter-day versions of Henry David Thoreau's blissful rural retreat in the forests of New England, producing not poetry but management or marketing reports in conditions of harmony with Nature. Our offices will be the Net, our commute a walk to the home-computer installation.

One can only be positive about the goal of reducing commuting. You, the reader, probably commute and probably wish you didn't; possibly you are even commuting as you read this. Commuting takes often at least an hour out of an employee's day, sometimes more. The costs to the individual in terms of time and health and to society in terms of environmental pollution, vehicle noise and aggravation are immense.

The majority of all office workers in the USA sit at a desk operating a computer or terminal. As high-capacity data networking becomes available in houses, all that is needed is to relocate that computer or terminal into the office workers' home and, hey presto, the commute has disappeared. Perhaps, admit the futurists, a weekly meeting at an office will be necessary to help provide social cohesion among the working team, but the great majority of daily commuter kilometres will have been eliminated.

To the proponents of this school, the benefits of telecommuting do not stop there. With Information Superhighways replacing the need for physical travel, the very *raison d'être* of a big city – the bringing together of people for daily commercial, professional and industrial contact – disappears. The phenomenon of high-density urban areas can become a thing of the past.

However, had the ability to stay in touch with colleagues over telecommunications networks been sufficient to release a typical office employee from the need to commute to work, we would have seen a very substantial implementation of this practice years ago. For many office workers, the functional requirements of communicating with colleagues were already available with the development of the telephone, the fax machine and rudimentary data interchange over modems, plus of course the availability of the personal computer. A massive shift could have begun at the end of the 1970s. Clearly it did not.

It is estimated that approximately seven million people in the United States, or some 5 per cent of the workforce, are telecommuting routinely. Over the past decade there has been enough accumulated experience to draw some initial conclusions, and particularly to know that the use of technology to reduce the rigidity of office attendance is generally far preferred by both employee and employer than a complete cutting away from the office. A 1996 article in *Newsweek*'s 'Millennium Notebook' cited 'the grandiose predictions of "travel" to HQ via the Information

Superhighway' but, surveying research in the field of workplace studies, concluded: 'Our flirtations with telecommuting have served mostly to accentuate what's useful about offices.'

Even though much more sophisticated technology is now available, nearly all full-time employees of companies still spend the working week on the company's premises. Employers will accept home working for freelances, temporary consultants and in isolated cases when, say, employees must largely work from home for a period because of the need to care for an infant. A very few progressive organizations are encouraging the habit of home working, either to promote the technology that makes this feasible or to single themselves out as a different category of employer. But the bosses of most top-performing organizations want their employees on the premises not just to conduct the functional processing of information but to take in the organization's values, be enthused and motivated by its competitive positioning, and generally get into the corporation's spirit.

In this regard corporate USA, Europe and Japan know a lot about human psychology in the workplace, and historical experience backs them up. Ever since Roman generals started to drill their troops on parade grounds, leaders have known that you can wring more performance out of individuals if you have their physical environment under control. Evidence from the birthplace of the Industrial Revolution, the Midlands of eighteenth-century England, shows that the gathering together of cotton-industry workers in factory-sized workplaces began as a means of obtaining greater productivity than had been possible under the previous contracting-out system. The arrival of machinery which required this concentration in one workplace for *technological* reasons came a considerable number of years later, and arguably the development of the technology came as a result of the grouping together in the workplace rather than vice versa.

If good telecommunications were sufficient to bring in attractive new organizational arrangements, most of us would already be telecommuting, teleconferencing and searching databases for optimal matching of buyers and sellers. Instead we are still crawling to offices in traffic jams and making our purchases by browsing through shops or in response to a personal sales call or a captivating commercial.

There is something elusive but obviously very powerful about human physical proximity, whether within or between organizations. Finance is the most heavily computerized of all sectors, with networked operations within banks and between banks and their customers. Yet three big cities – New York, Tokyo and London – dominate the industry as never before. A 1995 survey in the *Economist* pointed out that half of the world's 100 largest banks and essentially all the large security firms are based in these three cities. Where a particular specialization exists elsewhere, it likewise tends to concentrate around a small number of locations, such as Chicago for derivatives trading and Boston for fund management. The survey concludes:

> Banking seems to be 'in the air' in financial capitals – and this remains true even though money can fly around the world in nanoseconds. Face-to-face contact is irreplaceable, partly because it promotes the trust that is essential to making deals, and partly because the informal exchange of ideas in such businesses is unpredictable. Someone who has spent a day in fruitless meetings might discover a vital nugget of information in a chance conversation by the coffee machine. Serendipitous proximity cannot be reproduced by fax machine or videoconferencing.

In her book *The Death of Distance* (1997) Frances Cairncross likewise draws attention to the phenomenon of the clustering of businesses, which can readily be traced back to medieval times – as evidenced by road names like Milk Street and Baker Street in historic towns. She hails the ability of modern telecommunications to transcend geographical barriers but, like others who have studied its impact carefully, emphasizes how much geographical proximity still matters. She cites the example of the US movie industry, which is hugely dominated by the single town of Hollywood. Ironically, the physical location which is associated more closely than any other with its local industry is Silicon Valley; if any group of entrepreneurs should understand how the heady air of success can pervade a geographical area it is precisely those who are developing the new information technologies.

There is also the vexed question of what the new technology will do to assist in enhancing prosperity. Hopes of a white-collar-based transformation of business performance as dramatic as that achieved by the Industrial Revolution must be downplayed. Office automation will help to sell cars and fill airline seats, but will not *manufacture* those cars or make those planes fly. We must be realistic about the fact that, while the new technology is as important as the old as a competitive tool, it does not compare as a contributor to overall economic output.

Until 1903 there was no powered flight. From then until 1970 each new generation of aircraft was hugely superior to the previous one, culminating in the Boeing 747, which is still the mainstay of intercontinental flight. The productive output of the industry, however measured, grew by multiples. In the 1970s, by which time all major airlines had a modern jet fleet, the technological competitive weapons became office-based: the reservation and ticketing system, improvements in passenger administration and loading management, and so on. These were important, but they were not revolutionary by the standards of what science and engineering had achieved in producing the planes in the first place.

The latest step in the deployment of technology for competitive advantage has been the airlines' loyalty (frequent-flier) schemes, which make great demands on computer technology and, fiercely supported by travellers, are a vital tool in the fight for customers. The loyalty-card concept has since swept much of the retail universe. In these latest developments, the disconnection between private competitive gain and global economic gain is complete; the rivals are playing out a zero-sum game – a game in which one player can gain only at the expense of another's loss, and no benefit is conferred on the whole. These are not technological advances – better planes, greater efficiency – that will filter on to benefit all.

To the executive responsible for information technology in the offices of a giant corporation like General Motors, the notion that failure to keep up with the latest in office automation may not be of prime importance must sound like heresy. Could our personnel department function without computerized records on each of our hundreds of thousands of staff? Could our sales staff maintain customer relationships without online access to dealers? Could the planning staff calculate the next five years' revenues without spread-sheet programs? The answer is yes, of course they could, just as they did in the past. The personnel records were in file drawers, the salespeople spoke to dealers on the phone or in person, and the MBAs in the planning department scribbled their calculations on the back of used stationery. To some degree the use of information technology in a modern company setting provides a kind of corporate hygiene factor – crisp cleanliness in the office, with the data sorted, filed and available on demand.

Consider the contrary scenario. Instead of taking machinery away from the staff in the *offices* of General Motors, go down to the factory floor and take it away from the operatives there. Give them hand tools, hammers and the like, and tell them to make cars!

While, in the Darwinian struggle for customers and staff, businesses must keep reasonably up to date, every business will have to keep up with what office technology has to offer, the improvement in the output of the economy as a whole will be limited by a fact so obvious that it is seldom said straight out. This is that over 80 per cent of the final output of the economy – of Gross National Product – is of a physical rather than informational character: it consists of housing, food and vehicles rather than banking or educational services.

This apparent 'materialism' – in a very literal sense of the term – of the consuming public has an understandable foundation, namely that information is a descriptor, and *will not transcend in value the things it is describing*. To inform, to describe, there must be something *to* describe. Our appetite for information products is satisfied by a quite modest expenditure of our available income. This statement is not a value judgement: it is an empirical fact, as revealed by the relative amounts people of all income levels are willing to spend on cars, houses and clothes versus books, computers and banking services. There is no evidence from observation of the wealthy that, as income grows, we become less 'materialist' and more 'idealist' in our preferences. Relative to their income, the wealthy actually spend proportionately less on such items as television, cinema, newspapers and books (how much *can* a millionaire spend on media, after all?), allocating instead a larger proportion of their budgets to buying more beautiful homes, luxury vehicles, travel and costly art or antiques.

It is a truth that since the time of Pythagoras – and especially since that of Plato – society has recognized the existence of a realm of pure ideas, worthy of respect or even worship in its own terms; and in the information age this notion has been given a boost in intellectual circles. It is too much, however, to expect the human race to undergo a wholesale switch to Platonic idealism just because technology has given more plentiful access to information.

Had it been the case that the bulk of consumers' spending was on products of an intrinsically 'informational' variety, such as television programming, books, educational materials and financial services, the prospects for a transformation of economic output by information technology would be self-evident. There is no debate over the fact that the computer age has transformed the delivery of banking services, for example, or that it will shortly replace the physical delivery of videos. However, most of us – I certainly include myself – will pay respect to ideas, as ideas, from a small corner of our wallets and, with the rest of our cash, will buy information, if at all, only if it will help us pursue more tangible goals.

The fact that consumers do not wish to spend more than a small part of their income on these products and services presents a major obstacle to the fulfilment of an information-technology-driven economic revolution. It means that the major contribution of this new industrial revolution, as of its precursor, will in due course have to be to help satisfy our pre-existing needs and desires, rather than to be a source of new aspirations. There will be much evolution in product design, clearly. Cars, for example, will have an increasing number of electronic gadgets, but they will still be recognizably and primarily cars, and we will still want to drive them along real, not information, highways.

★

It was not clear at the beginning of the microprocessor revolution on which of the two great halves of the economy – the information-processing part or the production part – the new technology would make its impact. The first chips which entered commercial service were embedded mainly in machines – typical applications ranged from machine tools to dishwashers. Electronics was a branch of engineering, and microprocessors were part of the arsenal of engineering tools. The picture we may have had of the future of information technology when micro-electronic technology first came out of the laboratories in the 1960s could easily have been that it was to be used to further the traditional goal of technological progress: to improve our ability to extract, process and mould to our requirements objects of the physical realm.

Things were to turn out differently. The proportion of information-technology hardware and software expenditure going into industrial processes – the real world of factories, chemical processes, the energy sector and transportation – as opposed to the clean and clinical information worlds of the office and home-entertainment system is now less than 10 per cent. The extent to which our 'real', as opposed to paper, economy is computer-operated is generally overestimated in the public mind. Terms like 'computer-aided manufacturing' disguise the fact that the computer-assistance is often limited to such organizational aspects as the coordination of supply materials; truly automated fabrication generally takes place only on highly structured factory production lines, which constitute a small part of the industrial economy and are not predominant even there.

Some care must be taken with definition here. In 1997 the world's manufacturers spent some $90 billion on information technology, an impressive figure which might seem to imply that a direct assault was thereby being made on productive output. The car- and aircraft-manufacturing industries have been particularly active in computerizing the design process, and, especially in the case of cars, of the manufacturing itself. To quote a 1997 review of this technology in *The Economist*:

> In the past few years certain industries have been simulating more and more aspects of production in these pleasure palaces of the imagination before carrying them out for real and hence – so they say – saving a great deal of time and money in the process.

Examples include, in safety testing, the crashing of 'virtual' cars rather than real ones (though real crashing is still mandatory at the end of the design process), thus saving up to $1 million for each wrecked prototype.

But a breakdown (provided by IBM) of this data shows a different picture: $11 billion of the $90 billion went on sales and marketing support; $57 billion, around two-thirds of the total, went on what is encouragingly called 'production systems' (however, the subcategories here are business management, resource planning and logistics, revealing that this again is office support of production rather than production proper); $21 billion was spent on design and development (that is, computers for the design offices) – this is certainly closer to the mark though still not exactly at the coal face. Only a very small percentage of the total, then, was actually spent on technology for the manufacturing process. As *The Economist* put it, even when the Nirvana of fully automated design has been attained, the engineers working in the virtual world of forms will have to emerge for 'the unpleasant, if necessary, duty of turning them into things tangible enough for people to buy'.

The stunningly successful software firms of the Silicon Valley culture – Lotus, Oracle, Netscape, and of course

Microsoft, further north in Seattle – make word-processing suites, spreadsheet programs, database packages and browsers. Their counterparts who are concerned with the computing and software embedded in industrial machines are minnows by comparison. There is no Pittsburgh Valley to which entrepreneurs tempt venture capitalists with plans for conquering the steel-mill or central-heating automation market. It is as if the big guns of the computer industry approached, metaphorically, the doors of corporate USA, saw to one side the din and soot of the workshops and chemical plants and to the other the clean and clinical information world of the office, and expressed a decided preference for the latter.

There is sound economic logic underlying their decision. Walk into the offices of, first, a corporation which makes clothing and then of one which smelts aluminium and you will find the activity remarkably similar: executives on the phone, assistants filing papers, accounts clerks calculating invoices, and so forth. Visually the two are so similar that you couldn't tell if the firm were producing T-shirts or crane parts. Go to the factory floor, though, and the production processes could hardly be more different.

This illustrates the huge task which the new technology faces when it grapples with that real world of physical production. It has to cope with the enormous variety of processes which go on 'out there' in factories, on construction sites and down mineshafts. In the manufacturing, extracting, energy and distribution industries every patch and corner has its own processes. Skills and secrets unique to that nook of industry are tucked away in the minds of engineers and in their specialist journals. For computer technology to be applied here, each of these countless arcane skills has to be mastered by the programmer before it can be simulated.

When dealing with white-collar environments, by contrast, technology enters a more uniform world. There is only so much you can do with the 10 digits, the 26 letters, the graphics and the audiovisual signals that make up human communication, either live or recorded. You can type a letter and send it. You can speak on the phone with one person or teleconference with several. You can pay a bill or decide not to pay it. It is small wonder that office technology looks so uniform: white machines for a white-collar world of pure information.

The new technologies, both hardware and software, have cut their teeth on the clean and uniform world of information sifting, and rightly so. It is the remarkable homogeneity of the information world – contrasted with the heterogeneity of the industrial and natural worlds – that has launched the information-technology industry on to the 'virtuous spiral' of rising quantities and falling prices which it has enjoyed for the past three decades. But virtuous spirals do not go on for ever. The time is coming when, like the pocket calculator before it, the desktop office-support system will have been 'done'. Every office worker will have one, and the industry will have to look elsewhere for something new to market.

In a steel mill, computers put to work in the accounts department can produce a cleaner layout on the invoices; to get cleaner burning from the chimneys you need to write software not for the office but for the inside of the furnaces. There are programmers writing just such software.

Looking inside a cement kiln, you could be describing a scene from Dante's *Inferno*. The temperature is at least 1,250°C, approaching the melting point of iron. A slurry of clay, limestone and sand mixed with iron ore enters at one end of a slanting cylinder, which turns over very slowly as the raw materials combine in the heat to form a complex array of chemical compounds. Get the temperature conditions exactly right and small stones called *clinker* will emerge at the bottom hours later. These are crushed into cement. Get the conditions wrong and the clinker will not form.

Cement kilns for a long time defied computer control. The slurry supply, which comes from natural sediment, varies in composition daily. Human operators develop, over the years, a feel for the right balance of furnace fuel, raw-material input and speed of kiln rotation. Attempts were made to improve efficiency through computer control, but they always failed. No one was able to write a program which coped with the complexities and vagaries of this unglamorous but hugely important industrial process. Until, that is, the development of fuzzy logic.

In the annals of twentieth-century innovations conceived in the West but commercialized in Japan, fuzzy logic deserves a prominent entry. It was the brainchild of Dr Lofti Zadeh, an electrical engineer who undertook his graduate studies at MIT and Columbia. In 1973 he wrote an article which suggested that, to control complex 'real world' systems like industrial furnaces, computers should use algorithms which are blurred rather than crisp. The article was read by a Danish engineer, Lauritz Peter Holmblad, who worked in a cement-making company.

Holmblad had an intuition that, in his industry, fuzzy logic might succeed where more conventional digital control had failed. Here was an approach to programming which allowed rules to be vague. Instead of saying 'The furnace is 10° too hot, so reduce the fuel supply by 5 per cent' it would make rules of the kind: 'Weigh up all the factors that should determine the optimal temperature and then reduce the fuel supply a bit if all in all they suggest things are getting too hot.' Very analog in feel. No single factor would have absolute priority, and none would be ignored altogether. No limits were hard and fast; none of the rules applied at all costs.

Together with Jens-Jorgen Østergaard, one of his former students at the Technical University in Copenhagen, Holmblad spent two years refining a fuzzy controller for his employer's kiln. Their efforts resulted in higher cement yield and lower fuel consumption. The paper Holmblad and Østergaard wrote in 1980 describing their success drew attention throughout the world, and the two pioneers have subsequently sold hundreds of fuzzy systems to cement companies around the world.

The amount of unburned waste gas which emerges from a factory chimney is dependent in a complex way on the manner in which the fuel–air mixture is input into the furnace chamber, the shape of the chamber, and the movement and temperature of gases at different points inside it. Modern fuel-burning installations in industrial environments use computer technology to help the fuel burn optimally and to produce the minimum possible quantity of polluting exhausts. Even family cars now have microprocessor-controlled fuel-injection systems; their counterparts in an industrial furnace are obviously more sophisticated. Polluting factories, inefficient energy use and substandard industrial production are problems to whose alleviation digital-information hardware and software will make a huge contribution. But chemical processes still not home territory for the computer industry.

'Fuzzy logic', *faaji*, has become a catchword in Japan. Japanese washing machines use it to decide, by using sensors to assess the weight, texture and dirtiness of the load, what washing cycle to use. The logic is fuzzy because the choice of washing cycle is not determined by a rigid decision tree of the type 'This load contains coloured fabric so it should be washed at 40°C' but allows various influences to affect the equation. If the sensors assess the clothes to be particularly dirty (the popular Matsushita model puts a light beam through the water to judge the number of dirt particles) the software will reason: 'Although I would normally wash at 40° this particular lot is so stained I will turn the temperature up another 5°.' The computation is based not on fixed categories – 'coloured fabric, therefore 40°' – but on a blend of influences: 'coloured fabric, therefore *normally* 40°, *but*, since it is *unusually* dirty, I'll turn it up *a little*.'

This is, of course, how a human being at home would set about the washing decision – indeed, there is something reassuringly human about fuzzy logic. That reassuring feature is that fuzzy logic is analog. *Normally*, *unusually*, *a little* . . . these are not the binary distinctions of digital computation. A fuzzy program works with gradations of intensity, giving greater or lesser weight to different factors (clothing type, quantity, weight).

Industrial and chemical processes have a rich complexity which requires great flexibility in any system of rules that is to control it. Traditional approaches to computing have had difficulty coping with this complexity. By eliminating the rigid adherence to digital logic, fuzzy computation is fundamentally suited to dealing with the diverse character of biological and industrial processes. Adjusting in a gradual way for the interaction of many factors is exactly what fuzzy logic is good at, and in the case of the washing machine the outcome is cleaner washing for less energy – a typical saving is 15 per cent. If fuzzy logic can save 15 per cent of energy use in an appliance as simple as a washing machine, it is not surprising that enormous benefits can flow from its use in industry. The major ones are, in fact, in industrial processes hidden from public view, like the cement plants in which it was pioneered.

Fuzzy logic is a mere speck on the canvas of the worldwide adoption of information technology. There are an estimated 1,000 programmers specializing in this field in

Japan, and a few hundred in the United States. It may live up to the expectations of its creator, Dr Zadeh, and become the dominant mode of electronic control of industrial systems in the next century. Or it may find application in only a few niches. What is significant about it is that it is a deliberate, and so far successful, attempt to bridge the gap between the clean information environment in which nearly all electronic technology is presently deployed and the 'messy' physical world around us.

A much more ambitious attempt to bridge this gap is nanotechnology: the use of machines to manipulate matter at the level of individual atoms.

Nanotechnology burst into the awareness of the general public in August 1991 with the publication of a full-colour photograph of the first 'atomic switch'. A tiny probe was placed with great precision just above a layer of atoms of the element xenon, the whole apparatus being at extremely low temperature to minimize thermal vibration. By varying the voltage on the probe it was possible at will to make a selected atom move up to the probe and back down to the xenon surface.

Until now the technology of material processing has operated at macro scales, in which matter can be treated as having continuously variable or analog properties. Digital technologies, so far used mainly for processing information rather than physical goods, is in fact ideally matched to the nanometric (atomic-level) manipulation of matter, and this will be the foundation of much technical processing in the coming century.

At the atomic level, the deep structure of the physical world is also substantially digital. The smallest unit of a material substance is a molecule, which in turn is made up of a discrete number and arrangement of atoms. This in turn is made up of a discrete number and arrangement of subatomic particles (protons, neutrons, electrons . . .). Energy is also quantized – divided into discrete units – in accordance with the orbital energies of electrons around atomic nuclei. Though matter is digitized and quantized thus, space and time are, to the best of our knowledge, analog; they appear to be infinitely divisible, rather than coming in discrete or quantized lumps.

Nanotechnology derives its name primarily from the prefix 'nano', meaning one-billionth, but more directly from the nanometre, the unit of length which is one-billionth of a metre (or one-millionth of a millimetre); the diameter of an atom is approximately one nanometre. On the nanometric scale matter can be considered as essentially digital. The smallest unit of substance of an element such as iron is a single atom which, like a bit of digital information, is either there or not there – there is no in-between. The subcomponents of the atom itself – protons, neutrons and electrons – in turn subdivide further into smaller subatomic particles; though, if you divide this far, the substance you're dealing with is no longer iron.

A welder connecting steel beams on a construction site is working in the analog realm: he cuts where he likes and the positioning of individual atoms is irrelevant to him. The nanotechnology engineer is fabricating objects at the level at which the matter is digital: his is a digital technology which will move one atom, or two atoms, but never one and a half.

Since the early 1990s the significance of the field has mushroomed. Atomic-sized 'trains' have shuttled along tracks whose lengths can be measured in terms of molecules, picking up atoms at one 'station' and depositing them at the next. The great excitement nanotechnology is generating stems from the fact that there is literally nothing you can't put together with ultimately clean precision if you can master the technique of making it up from its individual atoms and molecules. A perfectly functioning automobile engine could be assembled from a lump of iron ore and a beef steak from supplies of carbon, hydrogen and oxygen (and a few other elements) – the steak need never have been part of a cow.

The ability to engineer material at atomic level was the culmination of several years' work by a group of pioneering thinkers working principally in the United States. The most prominent member of this group was Eric Drexler, a maverick scientist-engineer who, by his late thirties, had become the intellectual father of this new field but had yet to be awarded a PhD by his university, the Massachusetts Institute of Technology, largely because MIT did not know in which department to categorize his research. (Was it physics? Was it digital electronics? Was it mechanical engineering? The answer to all of these was, in a sense, yes and no.) Japanese finance provided $200 million for a research centre headed by Drexler to be created in Tokyo.

A vast chasm of development effort separates the atom-sized shuttle trains from the production of the cowless steak. Nevertheless, fundamental techniques – picking up individual atoms and attaching them as required to other atoms – have been mastered.

Practical applications are just beginning to emerge. The Sandia National Laboratory in Albuquerque, New Mexico, announced in 1998 the development of a nanoscale engine: though just the size of a grain of sand it is able to pull an object weighing half a kilogram. (The project team, which is funded by the US military, expects the first application to be an almost invisible locking device for a nuclear warhead.) Sets of transmission gears only microns across are produced by etching nanomachine designs into silicon wafers and removing unwanted material. The technique is reminiscent of the photolithography method by which logic gate designs are etched into silicon when making a microprocessor, except that in this case there are moving parts, and the resulting machine moves not bits but real objects; by turning cascades of gears with a cumulative ratio of three million to one it can pull, very slowly, that half-kilogram object.

The most revolutionary implications of the engineering of individual atoms and molecules are in the field of genetics. In May 1998 President Clinton, on a visit to Britain, dropped into a pub and surprised a retired couple by sitting at their table for a glass of beer and a chat. It was spotted by the press that, when he left, a security agent quietly appropriated the President's empty glass. We leave organic detritus behind us wherever we go, and hence voluminous clues, in the form of DNA molecules, to our hereditary traits, vulnerability to diseases, and much else. It is routine for steps to be taken to avoid foreign agencies getting hold of the President's code.

Science is beginning to understand and to be able to engineer another category of atoms: the individual component clusters of atoms which make up human DNA. Members of the public are variously excited by these developments and terrified by them; either way, they realize instinctively that this is the next century's equivalent of the nuclear bomb – and more. When news burst of the 'creation' of Dolly the sheep, the first mammal born as a genetically identical copy of an adult 'parent', political leaders in many countries, including the USA and the UK, set up commissions to examine the implications.

In a grey industrial building on the outskirts of the city of St Louis, Missouri, 200 researchers are leading a worldwide effort to identify the entire sequence of the human DNA code. Known as the Human Genome Project, the effort was launched in 1990; it has a $200 million annual budget appropriation from Congress.

DNA, deoxyribonucleic acid, is the substance in our chromosomes which contains the blueprint from which every cell in our bodies derives its design. Each molecule of

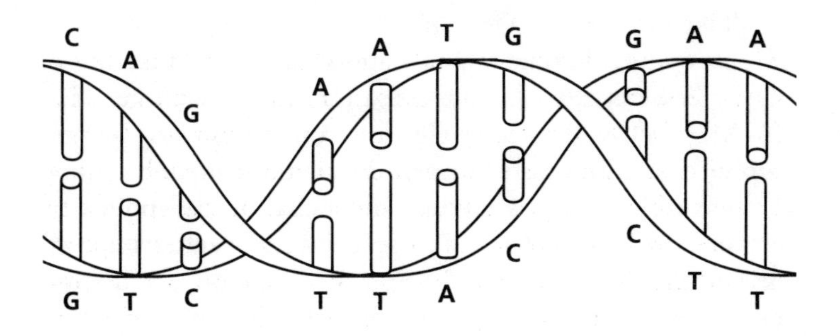

The DNA code of As, Cs, Gs and Ts in a human embryo is of absolute digital precision. But the cells in the embryo develop under the variable (analog) influences of their surroundings, making for an infinity of possibilities for the child that emerges.

DNA consists of a long chain – actually two chains coiled together in the now famous double-helix configuration; it is made up of millions of subunits, each containing one of four *bases*: adenine (A), guanine (G), thymine (T) and cytosine (C).

As the body of a growing organism develops, the protein molecules which will make up each new cell are guided in their formation by taking cues from lengths of DNA in a pre-existing cell. Depending upon what combination of A, G, T and C bases appear along a particular length of DNA – a gene – the protein molecules will be folded into different shapes. Your facial features are determined by the formation the proteins in billions of your cells folded into after they took their molecular cues from your chromosomes. Those features carry a family resemblance, more like your mother's or your father's depending on which parent's genes you chanced to pick up in the part of the genetic string that guides facial characteristics.

When the Human Genome Project is complete, science will have identified the sequence in which each of the three billion bases appears in our cells. Having a blueprint of a DNA strand which is 'right' is seen as a key step in identifying and possibly correcting genetic structures which are 'wrong' – those structures that lead to deformity or disease. There is a long way to go between identification of the complete genetic code and practical success in remedying genetic defects, but already individual genes within the complete DNA sequence (there are about 100,000 of them) have been identified and the genetic defect associated with particular diseases (such as Alzheimer's) has been identified and, very occasionally, corrected. The identification of the complete human genetic code will not be the end of the story, but it will be a very important symbolic – and hopefully also practical – landmark.

In parallel with the understanding of this biological code is the more dramatic prospect of altering it. This rapidly evolving technology is the biological equivalent of nano-engineering: it is the manipulation of individual atomic structures. It is a quintessentially digital technique, working on exactly identified groups of atoms and replacing them by other groups at the precise point required along the molecular structure.

The large-scale alteration of human genetic material would mark the beginning of an entirely fresh chapter in history. A 1995 survey of biotechnology and genetics in *The Economist* put it this way:

> The natural world, including the human body and mind, will become malleable. Implanted organs may refashion the brain, designer viruses rebuild old tissue. Human organs grown in animals for transplant are already being

> designed. New types of creature may appear; creatures to marvel at. If humanity can find no peers among the stars, it could create new intelligences on earth. All this will be made possible by genetics. Biological information will be stored in minds and computers as well as in genes. What was once unique to genes is now in humanity's grip. The control of biological information on this scale – of the raw data and the way it is processed – means the control of biology, of life itself.

Heady words, but in this field the rate of progress continually surprises. The task of sequencing the complete genome may be completed within a few years, earlier than the current target date of 2005. The biggest manufacturer of genetic-sequencing instruments, Perkin-Elmer, announced in May 1998 that it was creating a joint-venture company with the private Institute for Genomic Research; the objective of this for-profit enterprise would be to start the sequencing project from scratch using more automated techniques.

If the new joint venture goes according to plan, by early in the next decade academic researchers will be able to buy access to the complete raw data as it is read off our chromosomes. Corporations with adequate budgets will choose more elaborate software packages to help their researchers work with the sequence. Molecular biology, until recently science fiction, is beginning to be 'business as usual', involving buyers and sellers, market segmentation and niches, corporate ventures and profit-and-loss planning.

The technology community accepts openly that its primary focus is not on the problems of smoking chimneys and malignant growths. Nicholas Negroponte ends his bestseller *Being Digital* (1995) with the words:

> Finding a cure for cancer and AIDS, or inventing a machine that can breathe our air and drink our oceans and excrete unpolluted forms of each are dreams that may or may not come about. Being digital is different. We are not waiting on any invention. It is here. It is now. It is almost genetic in nature, in that each generation will become more digital than the preceding one. Nothing could make me happier.

A televised debate held in Britain on New Year's Day 1998 was one of many futurist discussions to be broadcast in the run-up to the new millennium. Commentators were asked to discuss what the next century might bring. The first to speak was from the UK's largest information-technology company, BT. He produced a dummy cat and placed it on his lap. Tomorrow's technology, he explained, would be like this, familiar and comforting, something you could talk to if you wanted computer assistance to come to your aid.

He was describing how you will be able to interact with the technology of the future as it analyses your needs and processes information accordingly. It will calculate the optimal temperature in your house, put on classical music, connect you on the phone to your mother, tell you how to fix the car or inform you what movie is playing on the television screen in the lounge. You need not be too precise in your instructions – perhaps you'll be able just to hint at what you want, or even let the computer (symbolized by the cat) take the decisions for you. The computer's learning-oriented program will have found out what sort of things you like to do at that time of day. Bill Gates has software of this kind in his spectacular high-tech mansion near Seattle; guests find that, as they walk through the house, music they have chosen follows them, light levels adjust, and views on wall-panels can be changed to taste.

There is nothing wrong with a technology being applied to what it is best at – and, just now, what electronic technology is best at is presenting us with the fruits of its power to manipulate information. When children play at a computer they escape into a marvellous fantasy realm. They can explore a zoo in virtual reality, or refight a historical or fictional battle. Whether accessing just the local CD-ROM or the World Wide Web, they are in that other world which we call cyberspace. The experience is fascinating and (often) educational. As adults we can also find cyberspace informative, entertaining and (often) productive.

Nevertheless, we shouldn't forget what it is that has given so many of us the opportunity to sit in domestic comfort in front of computer screens rather than living the pre-industrial life which Thomas Hobbes described as 'poor, nasty, brutish and short'. It is mastery over the material world. Combine harvesters garner our food, mills spin our clothing fabrics and machine tools fabricate our appliances. Antibiotics and surgery save our lives.

This mastery is still woefully incomplete. In developed countries we have food and material possessions in quantity, but are increasingly aware of the shortcomings in quality. For billions in the Third World there is not even the requisite quantity. Medical advances have been considerable, but ill health still afflicts all of us some of the time and some of us all of the time. Finally, such techniques as we have developed for mastering matter are hugely wasteful of resources and damaging to the environment.

In the past twelve months the population of the world increased by a further 100 million, and this growth rate is barely abating. Maintaining this growing population (with its growing expectations) in a state of even adequate well-being on the finite surface of the earth through the twenty-first century will require all the technical help we can muster.

Without doubt, at a time further into the future than we are here looking, digital technology will, for better or worse, come deeply to grips with the biology and chemistry of the world around us. The engineers, scientists, medical researchers and other professionals working on the chemistry of industrial processes, manipulating atoms using nanotechnology or mapping out the sequence of the three billion-long code in the human DNA molecule are taking the first steps towards a technological revolution which is likely to be wholly more transformative than any which will be brought about by the current generation of information products.

9

Who Are We in the Digital Age?

In the last months of the year 1899 the world's newspapers made predictions about the forthcoming century. The prevailing view, as typified by this commentary in Paris *Figaro*, was that the world was finally entering a period of scientific rationalism:

> It will be a majestic spectacle, the onset of which I should like to see myself. We shall begin to hope that the nineteenth century which nurtured us will sink into centennial oblivion together with the stupid hatred and all those idiotic mutual recriminations and cretinous calumnies which have darkened its last days and are unworthy of reasonable people.

The guiding force for this optimism was scientific progress. The explosion of technological knowhow at the turn of the last century, dramatic even by today's standards, seemed destined to end so many of the problems which had bedevilled mankind in the past. For this commentator – requoted recently in *History Today* in an article aptly named 'How the Future Didn't Happen' – and his contemporaries, there seemed not the tiniest sign that the advanced nations of Europe, of which they were the proud citizens, were about to embark on a half-century of strife awful beyond their imagination. The rate of technological progress would not disappoint – but it would be accompanied by appalling suffering in trench warfare, holocausts, gulags and killing fields as nations and ideologies entered into 'mutual recriminations and cretinous calumnies' as never before. It was not enough for there to be a rational new technology-led route forward. Much stronger forces – human greed, nationalistic, ideological and religious fervour, lust for power – were to play themselves out on the world's stage.

Maybe it is because another century is beginning; more likely it is because for the second time we are going through a period of exceptional technical progress – for whatever reason, the business of futurology is experiencing another bout of technological determinism. This time round it is the computer which is seen as driving us towards a more rational future, with processing power rising and costs falling as the chip motors along the learning curve. In 1998 Michio Kaku of the City University of New York, in his book *Visions* (reflecting his interviews with over 150 scientists, many of them Nobel Laureates) echoed the words of *Figaro* in 1899 when he concluded that

> the information revolution is creating global links on a scale unparalleled in human history, tearing down petty, parochial interests, building and forging a common planetary culture.

Wired magazine devoted a cover to 'The Long Boom', announcing:

> We're facing 25 years of prosperity, freedom and a better environment for the whole world. You got a problem with that?

The detail was contained in a chart entitled 'The Future History of the World' that described how technology would affect economics, politics and society, finally bringing about, some fifteen years from now, 'the beginnings of a global civilization of civilizations'.

In the 1930s the French scientist, sociologist and theologian Pierre Teilhard de Chardin predicted the emergence of a noosphere, a network linking mankind at the mental rather than the physical level. Teilhard described this noosphere partly in physical terms, as an information network, and partly in spiritual and philosophical language, as a force which would act to unify society. One of the many metaphors which he used to put the concept across was that of a 'halo of thinking energy' encircling the planet. Today the same combination of technical, sociological and philosophical terminology is used to describe the Internet.

Arguably, the technology – that is, the hardware – for Teilhard's vision is about to be with us. But when he spoke of a future noosphere he did not mean only, or even mainly, the means of physically reaching across the globe. He referred to a further development in the process of human evolution which would lead eventually to the attainment by mankind of a greater unity of mind, body and spirit; there would be a sharing of purposes and ideas and values across societies. When speaking of a future networked society, a global culture or a new phase of civilization based on the interchange of ideas, we must take care to distinguish between a network in the technical sense and a network in the human one. Technically, we have wired the planet. Yet at another level, that of our complex interaction in social and economic settings, much less is changing.

I have gleaned two lessons from the history, short as it has been, of electronic technology; they are, inevitably, over-simplifications, but they have been my guide. The first is to regard almost any prediction of the future power of the technology itself as understated. The second is to regard almost any prediction of what it will do to our everyday lives as overstated. We must keep in mind, as so often when analysing the impact of information technology, that this is a technology of machines which transmit and process data, and that the basic dynamic within and between organizations is that which takes place among people who originate and consume knowledge. If the analysis set out in this book is correct, computing machines will transform the channels but not the content – the form but not the substance – of our interaction with one another.

The new technology can be thought of as a digitally formatted stage on which individuals who work with knowledge will continue to play the parts which they, variously, already play. The engineers will engineer, the marketing executives will market, the bankers will bank, the teachers will teach, and so on. For the technology industry, the task is to continue to improve this stage, this infrastructure of machines and networks, in such a way that anyone of reasonable ability can act on it. For individuals and organizations in any field of endeavour, the challenge ahead is to continue, through investment in human capital, to develop the ability to play out those roles – roles in a drama which will be acted out between people, not between machines.

Lurking behind predictive scenarios of a computer-driven society is an emaciated view of what it is to be human: a model of the person as an entity whose objectives we have understood and can deliver by programming machines – who is satisfied with the simulation as much as with the reality, who is responding to images and sounds and not to the hearts and minds of those behind the images. In fact, we are exceedingly complex organisms, pursuing goals we don't understand, in ways which we cannot grasp, for reasons which we certainly have no ability properly to fathom. It is precisely these qualities of unfathomability and unpredictability – free will – rather than our powers of logic that make us still unique in the age of digital machines. There is no scientific evidence or philosophical argument that technology is developing in a direction to emulate those qualities.

The first round of the computer revolution was the introduction of mainframe computers to replace the armies of clerical workers undertaking routine data-processing tasks. Today a $1,000 notebook with a Pentium chip running Office 97 can look after an executive's data needs more effectively than a $2 million IBM 360 with its software-support team, and there is as little nostalgia for those big old machines as for the factory-like offices they replaced.

But in one respect at least the earlier era could be said to have been a golden age of computers. Precisely because they were so expensive and difficult to program, they were used for what they were suited to best and people suited to least: calculating, sifting and storing in a context, like bank-account reconciliation, where there is not the slightest advantage to be gained from the human/analog touch. They were used, in short, for computing.

Thirty years later, by contrast, digital machines could replace receptionists and answer the telephone – inspiring the Dilbert cartoon of a machine-generated voice saying to customers: 'Your call is important to us. Please hold the line while we ignore it.' Technology by this time was clearly beginning to enter into roles in which a human rather than a computational touch did indeed have value. Whether the bank would give you an overdraft became no longer a decision that would be handled by a bank officer but instead would be handed over to an expert-system program which would weigh up your creditworthiness by algorithm. The computer was surreptitiously crossing the barrier from what it was ideal for – keeping track of your bank balance and calculating the interest – to what it was not, because it was being asked to deal with problems which it could encode only by narrowing the problem to fit its own constraints.

Our societies are now embarking on the long task of integrating computing machines into daily lives. There will be many opportunities to substitute the quick decision of a machine for the more cumbersome human-generated thought process which preceded it. Each time there will be a small gain, direct and easily measured. But there are also costs being incurred – maybe small ones that are greatly outweighed by the benefit, but costs nevertheless. The person who is being replaced is not just a data-manipulating device, a preprogrammed Turing machine. It is well to remember that the substitution cannot be complete – to remember, in short, the role of people in the digital era.

The computer may have cracked the technical problem of encoding information digitally, but it cannot yet deal with us on a level to which we relate. In January 1998 Douglas Rushkoff wrote in the *International Herald Tribune*:

> We've bought the notion that our computers have brought us into the Information Age and that we are now in the realm of bits. It's not bits we're exchanging, but our very essence in the form of ideas, e-mail, graphics and chat. When I go online to engage in human interaction, I log off energized. When I search databases or shop in online malls, I leave the experience feeling drained and

> alone. This is because the former involves communicating with other living beings while the latter concerns only machines and their information.

To even the most avid fan of the Internet, the machines in the Net do not understand the terms and requests keyed in. The network's search engines will do their best and give you many leads to sites relevant to your quest, but still won't capture what you are *really* after. You are left with the deeply felt sense that the software is not communicating with you on your terms.

Virtual-reality pioneer Jaron Lanier is well placed to understand the power of the new machines. He puts it this way:

> Trying to understand how communication can be possible in the first place gets us into very mysterious territory. You can believe that a conversation between two people consists of objectifiable pieces of information that are transmitted from one to the other and decoded by algorithms, or that meaning is something more mysterious than that, something that no one has yet been able to find a method of reducing. I am in the latter category. I think that the fundamental process of conversation is one of the great miracles of Nature, that two people communicating with each other is an extraordinary phenomenon that has so far defied all attempts to capture it. There have been attempts made in many different disciplines – in cognitive science, in linguistics, in social theory – and no one has really made much progress. Communicating with another person remains an essentially mystical act.

The text of a poem or business memorandum is easily digitized. However, that does not make the poem or memorandum the product of a digital machine. It is the product of a human being; it is very humanly crafted and it requires a human mind to understand what it is communicating. There is a yawning gap between data shuffling, on the one hand, and thinking, on the other. We are certainly digitizing practically all forms of recorded information, and then processing them using digital techniques. But that does not mean that the information itself is intrinsically digital or, more importantly, that the operations which we as human beings wish to have performed on it are amenable to digital logic.

As the world becomes comprehensively wired we will need more than ever to understand the difference between data, information and knowledge – that raw data is a mass of symbols, that information is something more useful distilled from the data, and that knowledge is a still higher level of meaning, as information enters the human creative process.

The difference between data and knowledge is like the difference between raw food and the nourishment we obtain by eating it. An intermediate step, like information, is the meal we prepare from the raw ingredients and serve on the plate. Data's role in the quest for knowledge, like raw food's role in the quest for nourishment, is the starting point: the vegetables, the grains and the livestock in the fields. We then make it into a meal, selecting the raw foodstuffs according to our needs and processing them for taste and digestibility. The processing may be light, as in cleaning a salad or squeezing a fruit juice, or heavy, as in making milk into butter or meat into stew. Then we eat the prepared food to obtain nourishment: energy to live and chemicals to renew and grow our bodily cells. Knowledge is like the nourishment we receive after taking in and digesting the meal.

As modern farm machinery helps us grow the food we need for nourishment, so networks now bring us the data we use to build up knowledge. Modern technology has brought to every locality almost limitless access to up-to-the-minute data, just as we now find fresh food from every corner of the world in our local supermarket. And the comparison extends also to processing. The computer in the living room can scan, select and process the data into carefully tailored information nuggets of our choosing, just as the appliances in the kitchen convert the supermarket food into meals ready to eat. But to convert that information into the electrical and chemical changes inside our heads which store knowledge? That, like the conversion of meals into nourishment, is beyond the reach of technology.

Nobody has yet replicated what happens to food when it gets inside the body. We are extremely complicated chemical factories which Nature has evolved over billions of years to convert the food we ingest into carbohydrates, sugars, fats and proteins which are needed in various parts of the organism. Wouldn't it be like a dream, for some at least, if technology could give us the nourishment directly, under our instruction? We could, for example, make the spare fat store itself where we wanted, not where the metabolism puts it. Our muscles would be re-energized the moment we desired them to be – no more tired feeling at work in midafternoon.

But it's not like that – we can't penetrate the metabolic process and take it over. And the brain is an even more complicated knowledge factory than the lower body is a metabolic factory. Nobody has been able to replicate what happens to information when it gets inside our brain. As to how the thinking process works, science is as stymied today as philosophy was 2,000 years ago.

The new technology has done to data processing what the old technology did to fabrication. It has introduced – at the levels of data and information, not of knowledge – mass production. Electronics has brought the cost of digitally processed data down to the cost of the software doing the processing – at the margin, practically zero. Operations which can be done by a computer, be they the solving of a tax computation or the creation of a digital movie character, become replicable and marketable. Processed information has become a commodity. The same programs available to us are available also to friends, neighbours, colleagues and competitors. And, like plastic ballpoint pens and other mass-produced gadgets of the industrial age, they lose value as they become available to all. Floppy disks and CD-ROMs collect dust on our desks and in the kids' playrooms. Each of them contains millions of bytes of so-carefully crafted software – piled up high through the economics of duplication.

We have spent centuries wanting more information, so it's astonishing that it has become so cheap. But then who could ever have imagined in the eighteenth century, when a simple manufactured article like a knife could cost a month's income, that machinery of the complexity of thousands of knives would be left to rust in back yards – dishwashers discarded because a new model had come along or even just because the old one did not fit in with the colour-scheme of the new kitchen?

What will retain value, both personally and professionally, in this age of machines that can conjure up and process information in limitless quantities is that which computers cannot produce – just as what had value during much of the twentieth century was what could not come off a mass-production line. The source of unique advantage, of value, lies elsewhere. It lies in the minds of the millions who are the ideas-creators of the post-industrial age.

It has to be accepted that the ideas we create may want for quality. Thomas Aquinas's *Summa Theologica* is, by common consent, the greatest intellectual achievement of medieval Europe. A comprehensive restatement of Christian values and mores based on the philosophy of Aristotle, it runs in at least one current edition to thirty-three volumes – some 8,000 sides of text. And it was written in a thirteenth-century monastery equipped with a library and quill pens. We should not expect anyone to produce a finer work tomorrow just because information technology has advanced so very considerably since the thirteenth century. But the ideas will certainly be there in greater *quantity* than in previous centuries. More people have time, intellectual preparation and now – thanks to the Net, computer and CD encyclopedias – unprecedented amounts of information to work on. Mass-produced digital devices will give us every technical assistance.

James Bailey, author of *After Thought: The Computer Challenge to Human Intelligence* (1996), points out that we have been taught since the time of the ancient Greeks that rational thought is the pinnacle of human mental achievement: 'We as a species made a decision at some point to define human uniqueness around our intelligence.' If by 'rational thought' is meant logical processing, it is a bad place to plant our flagpole. Human intelligence is much broader. Schopenhauer emphasized almost everything else in our mental armoury: the power of will, drive, emotion. These are deeper planes of human interaction, and much more uniquely ours than data sorting. If computers have devalued one particular aspect of our intellectual powers – pure computation – by cheapening it, as industrial machinery devalued the ability to pull a heavy load, so be it. There are plenty of human qualities left untouched, and they will become the more valued. Bailey points out that, if the emergence of powerful computers like Deep Blue forces us to realize that logical processing is not a unique quality of humanity, then this is for the better:

> It's going to be a painful process, but if in that process we come to understand that we are not essentially analytical beings, that our essence is something higher, then that's a positive development.

What the future will bring by way of new combinations of human and machine information processing we cannot know. But so far the front-running hypothesis is that the creative engine remains the human, with the machine as its servant at a fairly low level of processing – excellent communications, good housekeeping of data, very hygienic in presentation. This hypothesis is also the one that gives us the most respect as people, and the one for which we should prepare.

It is not being Luddite or otherwise unenthusiastic about the possibilities of digital technology as a tool to refute the new value systems and philosophies surrounding it, or the notion that the technology will bring about a new phase of human civilization. There is a middle ground, one which recognizes the power of the tool while studiously avoiding the pitfall of assuming that the rationality of the new machines will be matched – whether for good or evil – by a rational approach to their use. The eminent Spanish–US philosopher Manuel Castells writes in the third and final volume of his *The Information Age*:

> The twenty-first century will not be a dark age. Neither will it deliver to most people the bounties promised by the most extraordinary technological revolution in history. Rather, it may well be characterized by informed bewilderment.

We do not yet live in the world depicted in Andrew Niccol's film *Gattaca* (1998), Hollywood's first full-scale exploration of a world in which genetic engineering is routine and babies' genes are tested and 'improved' within days of conception. But, as knowledge accumulates exponentially, new classes of issues will arise which have never faced humanity before. Biotechnology, especially genetic engineering, could change our species; the Institute for the Future has held a meeting, rather chillingly, on the coming of the 'post-human' society. In the next half-century there will be shifts in economic and social structures and in individual values which we cannot predict but which will be as transformative as any we have so far experienced – tinkering with human DNA might make this an understatement. The coming decades will bring surprises to leave us gasping – developments which will make the electronic revolution taking place in our offices and homes seem minor by comparison.

The strategic and military consequences of the path this technology is taking are not, in the main, weighing on the minds of its developers.

Columnist Thomas L. Friedman, writing in the *New York Times*, said of the present culture in the industry:

> There is no geography in Silicon Valley, or geopolitics, only stock options and electrons. Unless you look at both geotechnology and geopolitics, you cannot explain (or sustain) this relatively stable moment in world history.

Robert Kagan of the Carnegie Endowment has made the same point in different words:

> The people in Silicon Valley think it's a virtue not to think about history because everything for them is about the future. But their ignorance of history leads them to ignore that this explosion of commerce and trade rests on a secure international system, which rests on those who

> have the power and the desire to see that system preserved.

The intense but somewhat single-minded culture born in computer laboratories in the 1970s – a drive for peer respect, more recently spiked by the taste of wealth – is still prevalent in the hotbed of creativity which is the high-technology industry. The checks and balances that will be needed to prevent chaos – a fanatical Third World group declaring war on the Information Superhighways of the West, perhaps – would come more suitably from outside that particular culture; but how many outside the culture really feel for the problem?

A press release issued by IBM in January 1998 describes a software package which allows the three-dimensional visualization of objects on a screen. It is being used to help reconstruct Dresden's Frauenkirche, one of the numerous buildings destroyed by the firebombing of the city in February 1944. Most of the historic edifice was reduced to a pile of sandstone rubble which has remained on the site ever since; to quote the computer firm – may the elders of Dresden forgive the ludicrous euphemism – 'Allied bombing changed the church's architectural make-up for ever'. Now the new software is being used to enable as much as possible of the original sandstone – some 30 per cent – to be put back in its rightful place during the reconstruction.

At basically optimistic times – times of economic boom, like ours, when people are looking to the future with more enthusiasm than foreboding – the presumption is that technology will drive change for the better. But for a brutal (and ironic) reminder of the strict neutrality of technical progress in the human drama, one need look no further than IBM's next press release on the same visualization software. It describes a different application, in a factory in Munich, where the package is 'trimming production costs, shortening throughput times and optimizing the process chain'. The products being manufactured? Guided missiles, tomorrow's bombs.

As the millennium approached, a much sterner lesson was being learned by society as it adopted the new technology, a reminder that information machines can impose on society large-scale economic as well as security risks when an error occurs.

Worries about the millennium bug have been circulating among computer experts since the early 1990s; by 1997 they were being analysed with concern in the information-technology departments of banks, utilities, processing plants and distribution companies. In early 1998 the sounds of panic began to reverberate around the boardrooms of the world. Estimates of the cost this tiny programming quirk might inflict reached many billions of dollars, then began to top a trillion. The hurried patter of feet abandoning a sinking ship could be heard; headhunters reported droves of IT executives leaving the jobs where the systems they had installed might be infected – to take up jobs running systems equally tainted, but where they could be part of the solution rather than the problem. Insurers wrote liability for millennium-bug losses out of their policies. Auditors demanded directors' declarations as to the adequacy of steps being taken to combat the ailment. Government authorities in the United States, Europe and the rest of the developed world set up task forces and allocated budgets to publicize, cajole and threaten business leaders to face the problem.

The millennium bug brought home the uncomfortable truth that the leaders of commerce and industry, so much in control of the other technology and practices within their organizations, had not been in control of their information technology. Not only did they not know whether the computers were adding to profits, they did not really know what the computers did; they certainly didn't know what would happen if the computers stopped doing whatever it was that they did.

This new technology, in the literal sense closer than any other to company management because it was a technology of the office, was, in the sense that actually mattered, more remote. The way in which the computers went about the silent task of processing the data which flooded into and out of them was in reality a mystery to the people in charge. They did not know what standards were being employed to check data, to document programs, or to decide when code should and should not be used. Corporations, therefore, had little idea what to do in response to this predictable yet unexpected threat, or indeed how serious it might be.

The various other technologies deployed in modern economies have been subject to checks and balances that have been finely honed over many generations. Factory and construction equipment is designed and installed by engineers who acquire their qualifications after five or even seven years of education and traineeship. The same is true of administrative practices: the financial operation of any firm is under the control of senior accountancy personnel who have graduated typically twenty years earlier and, as they made their way up through the profession, gathered experience relevant to the ways in which financial controls can work or fail.

Software writing, however, is a profession that lacks such checks and balances. The fact that, in this new technology, the professional path is not one of apprenticeship and careful graduation from less to more important tasks as individuals broaden their experience of life is both a source of dynamism and a potential danger. At 18 you can be a professional programmer; at 20 you can be an old hand; and at 22 you can be at the top of your profession. Nothing wrong with that if the deployment of the programs is governed by procedures and professional practices which have been geared to deal with the unexpected, when there is supervision by people who have seen what happens when machinery fails or humans err. But, as the affair of the millennium bug has shown most vividly, software deployment is not under any such control. Among IT professionals there are many eminently responsible people, but the culture is still very much one of the loner, the creative genius, rather than the safety-conscious bridge designer or the cautious and sceptical financial auditor.

The millennium bug has, then, sounded a warning bell, and not just to firms but to society at large. It offers a graphic example of the information-technology project getting slightly ahead of itself. Not to the point of disaster – the billions or even trillions of dollars spent to correct it is a cost which the global economy can, if somewhat unhappily, take in its stride – but enough to make it worth asking whether the information-technology mores should be tweaked to give more weight to reliability, care and checking and less to immediate performance.

In fact, the bell was sounded at a good time, because the problem hit the world while digital technology was still largely a white-collar affair, limiting the chaos to the administrative and financial side of commerce and industry. The consequences of, say, a supermarket chain's automated stock management system deciding on 1 January 2000 that it is 100 years behind in fruit stocks, is not that its stores will actually find 100 years' worth of fruit being delivered through their doors. Such systems are currently automated only to the point that computer-generated *instructions* are given – to drivers, warehouse managers and other human operatives, people who would not be so stupid as to implement the absurdity. Consider, by contrast, how much greater the alarm would have been if the trillions of dollars' worth of computers in the world had been embedded in the 'real' economy rather than the 'information' one, so that trucks and other industrial plant started blindly implementing 100 years of activity in a single day. Truly awful chaos could have ensued, the stuff of a Hollywood horror movie.

The phase of the digital revolution which the bug affected was the first phase: the widespread adoption of computers by businesses. There are at least two more phases to go, each rendering us vulnerable to damage much more costly than that wrought by the bug.

The second phase is networking, which began to be adopted extensively in the 1990s, in the form of the Internet, but which will clearly have its major impact in the coming decade and beyond. Linking the world's information sources by so powerful a communications system provides enormous opportunities for economic efficiency and social connectivity – but brings a completely new level of danger from abuse or failure. At least the millennium-bug problem can be addressed by each major computer installation or group of installations independently: the computer systems of companies and other organizations are still primarily self-standing. The linkages between them are a secondary feature, a later add-on.

With the arrival of the networked world this is changing. Networks are what computing is to be about; it is precisely through the comprehensive interconnection of computer systems – of companies with customers, of governments with diplomats and the military, of families in one country with relations across the globe – that we will arrive at the next plateau on the continuing path towards an information society.

Along with the benefits will come unprecedented opportunities for damage through error and ill will. Tasters of what awaits us have included the success a 13-year-old in a small town in England had in penetrating the deepest secrets of the Pentagon in Washington. Congress has responded to this episode with a penetrating investigation, but will be powerless to avoid a more dangerous repetition as networks continue to integrate. The standardization offered by Internet protocols is a boost to hackers, spies and terrorists, making easier the task of roaming the world's wires in search of the computer installations that enemies need to access to wreak their havoc.

And what of the third phase, the transition of computers out of the information environments of offices and living rooms and into the 'real world' of genes and atoms? The imagination can hardly begin to grasp what the consequences of error or fanaticism could be. By the time computers are mastering matter at this *real* level, rather than an administrative one, it would be well for us to have improved our mastery over them.

We must make a choice as to how we are to prepare the next generation for this changing and no doubt bewildering world. A widely held view is that the most important new factor in education must be the new technology. In a 1996 poll US teachers ranked computer skills and media technology as 'more essential' than biology, chemistry and physics; than European history; than reading modern US writers like Steinbeck and Hemingway, or classic ones like Plato and Shakespeare; or than dealing with social problems such as drugs and family breakdown. Reporting these results in *Atlantic Monthly* Todd Oppenheimer described actions being taken by schools across the USA in response to such views. In Mansfield, Massachusetts, administrators dropped proposed teaching positions in art, music and physical education and spent a third of a million dollars on computers. The President's National Information Infrastructure Advisory Council, a group of thirty-six leaders from industry, education and other fields, provided endorsement for this approach at federal level. The council has suggested reducing various activities in order to make way for computers, among them field trips – direct experience takes second place to the screen.

In 1974 an infant Italian boy, only days old, suffered an injury to one of his eyes. As a routine part of the treatment, a patch was put over it for the duration of the two-week healing period. When the patch was removed he was blind in that eye and, sadly, he will stay that way for the rest of his life. Because of the patch his brain detected no light signals coming from the optic nerve, decided the eye was dysfunctional, and so failed to grow the necessary connections into the visual cortex. That mistake will not be made again. With each new such case medical scientists learn more about the remarkable malleability of the young brain. The way it wires itself up is enormously dependent on the early stimuli it receives.

By the time children are of school age they do not risk blindness in the eye being caused by lack of optical stimuli. But they do risk blindness to ideas and to ways of thinking, analytical skills and creativity. If we try hard enough we may indeed produce a generation which is more at home with data than with knowledge, with numbers than with cultures, with computers than with people. We can aim education so that the new generation will feel more comfortable with the analytical thinking in which computers can assist us than with the softer modes of thinking which until now have been afforded their due place in education.

But would that not be the worst possible outcome? Computer specialists speak as if their non-technical counterparts underestimate the significance of the information revolution. My concern is the contrary – that it is these experts who may underestimate it, by focusing on the digital bits in machines. What is happening is much broader, a technology-assisted speeding up of the surge of ideas which began with the Enlightenment and is taking us to . . . we don't know where. To cope with it we will need all the mental flexibility we can build. That non-digital quality of wisdom, and the creativity of which we seem to be capable – those are the assets which will help us tackle the problems of the twenty-first century. As information technology becomes more pervasive in our lives, it is easy to overlook how important the softer side of human reasoning was – and still is.

During the last technological transformation of comparable scale and speed – the one which between 1875 and 1950, the span of a single lifetime, moved us from an essentially pre-industrial life to electricity, cars, jet travel, domestic appliances, steel, plastics, modern medicine and the nuclear bomb – education retained its emphasis on fundamentals, not practical training. The world was awash with new techniques and products, but schools did not put these on the syllabus. Although industrialization was proceeding at enormous speed, the educationalists did not drop algebra or literature to teach car mechanics or domestic wiring. The stress remained on mental training through mathematics and sciences, and on understanding the totality of the human experience through the humanities. Able people became industrial and commercial leaders without learning at school the specifics of the technology their industry or commerce employed. They achieved their successes because their minds were acute, creative, sensitive and suited to solving problems in changing circumstances.

Historically, civilizations have maintained a balance between the emphasis given to the logical and precise modes of digital reasoning and the softer and more humanist – analog – view of the world. University students majored, in more or less equal numbers, in the humanities and the sciences. It should not be assumed that technology has overturned the age-old balance between the two modes of thinking: the vigour of mathematical studies and the less easily definable understanding of human nature learned through history and literature. The penetration of computer technology does not imply that more of the prizes of life will be handed to those who have mastered the logic of digital machines.

With human knowledge accumulating at such a pace, the prudent are limiting themselves to one forecast only: change. In the next half-century there may be shifts in economic and social structures enabled by this technology, or by a different one, or just through the thrust of societal forces, as transformative as any we have experienced. If this brief history of the cyber age is correct – if information technology is playing only a part, as follower more than leader, in a much more comprehensive knowledge revolution – the last thing the next generation needs is to swap its education in the lasting richness and complexity of the human spirit for expertise in today's computer technology. The future will belong to those who participate in the new information age in its broadest sense, not in a narrow one.

It would be too simplistic to dismiss the current vogue for seeing the future of human affairs as dominated by the computer as a repeat of earlier periods of technological determinism. The technological elite is, at its best, keenly aware of past mistakes in this regard. That elite is, nevertheless, convinced that this time around things are different. Computers really get to us in a deeper way: the idea has gained ground that there is something of us about them.

Bill Gates put these ideas across to an interviewer from *Time* in January 1997:

> I don't think there is anything unique about human intelligence. All the neurons in the brain that make up perceptions and emotions operate in a binary fashion. We can someday replicate that on a machine. Eventually we'll

> be able to sequence the human genome and replicate how Nature did intelligence in a carbon-based system.

His interviewer pressed him: wasn't there something special, perhaps even divine, about the human soul? 'I don't have any evidence on that,' Gates replied.

Such comments on human and machine intelligence reveal a thoroughly computer-like view of what constitutes human biological life and consciousness. The human genome, the three billion-long combination of atoms inside each of our cells which allows us to replicate ourselves so accurately, is tantalizingly digital in its precision. The exact sequence in which this DNA molecular sequence is arranged has led to excitement in the computer community that there is something deeply digital – and hence computer-like – about human life. The notion that the interchangeability of people and computers is deeply grounded *philosophically* would tie in nicely with the prognostications of the digital age. Its most energetic protagonist, Richard Dawkins, wrote in *River Out of Eden* (1995):

> Francis Crick and James Watson should, I believe, be honoured for as many centuries as Plato and Aristotle. What is truly revolutionary about molecular biology in the post-Watson–Crick era is that it has become digital. This digital revolution at the very core of life has dealt the final, killing blow to the belief that living material is deeply distinct from nonliving material. There is no spirit-driven life force, no throbbing, heaving, pullulating, protoplasmic, mystic jelly. Life is just bytes and bytes and bytes of digital information.

But it's not that easy. There is no denying the importance of the DNA discovery; it is clearly the defining discovery of the twentieth century in biology, which looks set to become the defining science of the twenty-first. It is true that DNA molecules, being made up of precisely arranged sequences of the four atomic combinations, lend themselves especially well to digital techniques of analysis. However, to read off the digital sequence of the DNA molecule is a far cry from understanding how this sequence translates into the remarkable phenomenon of life.

The way digital genetic code maps itself into actual biological shapes, structures and operations is still one of science's mysteries, and will remain so long after the sequence itself has been fully read off. The mystery arises because of the great intricacy of the inside of a cell. Although your body consists of a trillion of them, each of the cells is itself a finely balanced chemical-engineering factory containing thousands of different types of molecules, all interacting among themselves. Even individual cells are, after all, complex enough to be capable of separate life and reproduction. The widely studied *E. coli* bacterium uses a genetic code of 50,000 digits to reproduce just itself – a single cell.

The folding of a protein inside this living chemical factory is not only guided by the DNA molecules; it is subject to numerous environmental factors. The consistency of shape produced by the digital code is impressive. Identical twins come from the same fertilized egg, and hence the placement of every atom in their three billion-long DNA sequence is the same; the visual similarity at birth can be remarkable. But, deep down, as in the way the brain formed, the differences produced by the folding process are there: studies show that there is as much variation in intelligence between pairs of genetically identical twins as in the population as a whole.

Each new life starts out with a digital code for reproduction but also develops very much within the analog fabric of the material world.

To say that human life is, in essence, digital code because our cells reproduce with the help of numerically precise arrangements of atoms in our genes is not a scientific statement but wordplay masquerading as philosophy. The inference that the body's functioning is computer-like (digital) because its molecules are made up of exact numbers of atoms is like concluding that ballroom dancing is digital because dancers have exactly two feet. What defines a dance is how the dancers move around the floor. And the dance of the molecules as a cell evolves under the guidance of its genes – the protein-folding process – is, as biologists know well, of quite awesome intricacy. It operates under a combination of chemical and other influences which are very much analog, and not remotely understood.

The very idea that we are completely described as six billion digits is nonsensical: that which is analog is by definition of infinite variability. Insofar as we wish to call the sequence of atoms in DNA *information*, this information can be used only within a context, a physical environment. And this context is the hugely complicated chemical factory that is the human cell. The information can be read off only within that 'throbbing, heaving, pullulating, protoplasmic, mystic jelly' which is a living cell. The genetic material is not in itself a living cell; it is part of the mechanism by which a *new* living cell is copied out of a *pre-existing* one. The information required to produce each newborn creature is much more than its DNA. Insofar as our bodies are constructed in accordance with an instruction set, it is not just the one in our genes but also the one that has been produced as a result of the cumulative effect of four billion years of protein development.

Think through a scenario. An exchange of neutron bombs is about to obliterate all life from the face of the earth – every microbe, every insect and, of course, every human being. A group of technicians makes a last gambit to give the human species another chance. They put together a nanotechnology robot which can assemble molecules of DNA, atom by atom, and leave it a CD containing the complete human genetic code. When the neutron radiation ebbs, the robot picks up the CD, reads off the sequences and reconstructs the spirals of genetic matter, exactly according to the coded specification. It winds them alongside each other in pairs, just as they are wound together in a human embryo cell which is about to reproduce, and waits for life to spring forth again from this precisely configured data-set.

And spring forth it will not.

Whatever it was that caused life to begin, back in prehistory, it will have to cause it again if plants and animals are ever to return. The agonizingly slow evolutionary process will have to begin all over once more – billions of years of thrashing around of increasingly complex proteins until, *maybe*, a living being emerges.

To reiterate, the fact that the atoms in protein molecules (such as DNA) are digitally precise does not mean that the molecules *are* just information, equivalent to bits in a computer program. Molecules make up the *real* world, not the virtual world. Information is a quality we as humans can interpret from things, but it is not the things themselves. I emphasize this distinction because the term 'digital' is, understandably, so widely associated with the term 'information' that the two can sometimes seem synonymous. In fact information has both analog and digital character – as in a vinyl record or a computer program – and so, with digital atoms in analog space and time, does matter. We do not need to descend to the atomic level to appreciate these distinctions, for they occur in everyday life. The number of chairs around a dining table is digital – you can place seven or eight, but not seven and a half. The placing may convey information – if there are seven chairs and eight people arrive for dinner, the implication is that someone arrived uninvited! – but nevertheless each chair is a chair, not an item of data.

In the year 2005, if the fifteen-year international project to achieve this goal runs to schedule, biologists will have sequenced the human genome; we will know in what sequence the three billion A, T, G and C molecular structures appear in human chromosomes. A few scientists (and more than a few journalists) will then announce that we have decoded our DNA. But what we will have done, in fact, is merely to list the order of the atoms on it. We will 'understand' our genetic code in the way that a child 'understands' a scientific article if he or she can identify the sequence of alphabetic letters in it. Bill Gates spoke of 'sequencing the human genome and replicating how Nature did intelligence'. The step from 'sequencing the genome' (making a digitized listing of atoms and transferring it to a CD) to understanding 'how Nature did intelligence' is akin to the step between listing the letters in an encyclopedia and understanding the wisdom contained within it: we're talking about two quite different things. Understanding 'how Nature did intelligence' is not a fifteen-year project: whether it will take 150 years or 1,500, or whether it will just never happen, is anybody's guess.

The digital revolution has certainly provided an intriguing new toolkit for thinking through questions at any level, even the philosophical. What can't be allowed to go by is the claim that, by deciding we're computers, we've cracked the mystery of human life. Through the ages people have spent their lives grappling with this deepest of puzzles, deploying in their quest philosophy and science and theology and history and literature and art, and they have cumulatively built up not any kind of answer but only a glimpse of the vastness of the question.

★

The afterlife will be like the Internet, said a priest in a recent broadcast on BBC Radio. We will interact, have experiences, gain knowledge – but we will not be there physically, because it is not a physical place. He was telling his audience that we can now comprehend how a soul can live on for ever, unconstrained by space or time. A soul, in his depiction, is noncorporeal and immortal, pure form without matter, like the code in a computer program. Deep down, we are pure knowledge without a place in time or space; the flesh is just its temporary incarnation. A nice analogy – the afterlife as cyberspace: disembodied intelligence, dematerialized existence. Arch-Darwinist Richard Dawkins, of course, has no time for clerics, nor for what he calls their 'mystical and obscurantist views of life'; it is the selfish gene, pure genetic data, which is the key to our existence. Like the priest, though, he uses digital data as the basis for his parable.

History has its grandest sweep as the history of ideas: the foundations of philosophy laid in ancient Greece, the mysticism of the Middle Ages, the scientific revolutions of Newton and Einstein. A new chapter in this history of ideas is being written, timed to perfection (even if by coincidence) for the new millennium: the existence of a digital world to rank alongside the natural one – bits in silicon to compete with life in carbon.

Plato thought the world was virtual, consisting of idealized Forms; his student Aristotle saw it as real, tangible Nature around us. Thinkers have been arguing ever since. Western philosophy, in the memorable summary by A.N. Whitehead, is a series of footnotes to Plato. By showing how intricate and rich can be the realm of pure Forms, digital processing has given the Platonic model an encouraging nudge forward.

In philosophy as in the business office, computing provides a new toolkit for working through old questions. But it does not take sides. The biggest questions still remain unanswered, and each of us we will have to continue to look for the answers, if we wish to, using whatever combination of reasoning and beliefs we find convince us.

The journalist interviewing Bill Gates for *Time* did not let up on his questioning after the interview ended. He e-mailed Gates on several technical issues but also on the big ones: Can intelligence somehow be replicated in binary code? Has watching a two-year-old daughter learn to smile at her father's face changed his view at all? Answers to the technical questions came back promptly, but on the deeper ones there was silence. Finally, weeks later, a message from Gates arrived in the e-mail system:

> Analytically, I would say Nature has done a good job making child-raising more pleasure than pain. But the experience goes beyond analytic description. Evolution is many orders of magnitude ahead of mankind today in creating a complex system. I don't think it's irreconcilable to say we will understand the human mind someday and explain it in software-like terms, and also say it is a creation that shouldn't be compared to software. Religion has come around to the view that even things that can be explained scientifically can have an underlying purpose that goes beyond the science. Even though I am not religious, the amazement and wonder that I have about the human mind is closer to religious awe than dispassionate analysis.

Could it be that even the standardbearer of the new technology hesitates to embrace the new idea that a human being is a digital machine when confronted with the living evidence of his own growing child? While the new computing machines are spectacular in their digital precision, there is a more mysterious information-processing device which resides in each of our heads. Any forecast of the future path of the information age must acknowledge the human brain's abundant patterns of thinking and feeling, the richness of this natural ability of ours, and the complexity of the social fabric which human minds have built up over the course of the millennia. Fifty years of computer technology will not substitute as easily as technology enthusiasts would have us think for the natural processes of interaction between people and other people, and between people and their surroundings.

Epilogue

Emma (her real name) is 11 years old. She lives during the week in Boston, Massachusetts, and spends the weekends in a farmstead at the base of Mount Washington, near the small town of Jefferson, New Hampshire. She likes drawing and writing poetry, the subject matter being invariably Nature; depictions of butterflies, trees and hills, accompanied by whimsical lines in a child's hand, are framed on the farm-house staircase. She attends ballet classes, which she enjoys, but her biggest passion is for horses. Riding and grooming are the chief lure of the weekend home, to which she is driven in a black BMW M3 of which she disapproves – 'It's a drug-dealer's car, Dad.'

Actually, what her father deals in is specialized business software, which accounts for the liberal sprinkling of computers on desks in both houses. On these computers Emma does her homework, plays games, surfs the Net and writes to friends and family.

One often hears it said that it is only today's children who will be truly adapted to the computer – who will have grown up completely at home with the new technology. Emma is certainly at home with it. This term her school is doing a special project on Africa, and her responsibility is Ghana. She is collecting materials on that nation's geography, its peoples, its economy and its political systems, from which she will put together a paper to present to the class. The computer is her research tool for preparing the project; it is also her typewriter for producing it and her postal system for delivering it – not to mention, of course, her game machine for diversion when she is fed up working on it.

The world has embarked on the long process of taking advantage of the tools which have been created by the new revolution in technology. The tools will be used to create many things that we do not want but, on balance, many more things that we do. In that regard this revolution is not different from the previous one, which transformed daily life in the late nineteenth and early twentieth centuries; indeed, it is an extension of it.

In many respects the new technology is more benign than the old. Its products are tiny and consume minimal resources; their production does not kill, pollute or cause cancer. In one respect, though, we must be more on our guard, not less. These new tools are *mental* tools, not physical ones, and, just as the previous technology had the power to change our physical world, this one has the potential to change our value system. This technology is better at processing data than knowledge, information than wisdom, facts than ideas. If we decide that we should adjust our sense of what is important to reflect what the technology can do, perhaps for convenience or in pursuit of efficiency, we will change our value system. By moving away from what is human to what is mechanical, we will take a step backwards.

At some level, any object can be described in numbers, in bits and bytes. Guides showing school parties around the great medieval churches of Europe reel off data like baseball commentators. Notre-Dame de Paris is 130 metres long and 48 metres wide. Its roof is 35 metres high, and its two massive early Gothic towers rise 68 metres above the square on which the church was built. Construction of the building lasted from 1163 to 1250 AD; further embellishments took another hundred years. The colossal statue of St Christopher stands 8.5 metres tall, and the slender spire, or *flèche*, on the roof is 30 metres high. And so it goes on . . .

But the numbers do not convey the spirit of the cathedral, the beliefs which launched its construction and the determination which brought it to completion. These qualities the visitors will have to comprehend for themselves, and if anything helps them do that it will be not the statistics but the artistry and the majesty of it all.

We are only one generation into the digital revolution, and a fascination with the new scale of availability of information is understandable. But we must not mistake gigabytes for wisdom, or megahertz of clock speed for intelligence.

There is a danger that the computer will be allowed to impoverish the human experience. Some overreaction to the potential of this new offering is inevitable, as it was to the offerings of the industrial age. In Emma's home town of Boston, a $15 billion federally subsidized project is now under way to put underground the unsightly elevated highways which had been allowed to blight the city centre in the heyday of the automobile; the reclaimed land will be restored to greenery. There will be some blighting of our mental landscape by the new technology, just as there was of our physical landscape by the old.

★

But, from the evidence of Emma and her classmates, the computer will be kept well in check. To her generation it is very much: a tool to get things done, not something bigger than themselves and their friends. Emma would be surprised to hear that learned men and women are debating the equivalence of humans and software programs, or that educators are saying that computer skills should displace history and geography in her school. The difference between these machines and people is more obvious to her and her contemporaries than to those who were already adult when the computer age burst upon them. Because the young have played with computers since the age of three, they see right through them. They see the graphic spectacles appearing on screens for what they are: artifices of a machine which may be impressive and fun to play with, but not like the real human life going on around them.

Emma is not especially interested in computers. She is getting to the age when she wants to know more about people, cultures and societies. The boys in her class dream of football stardom and cars, but not, in the main, of computers. Those who spend their free time by the screen rather than outdoors are seen as inadequate rather than impressive, escaping from the rough and tumble of life rather than getting to grips with it.

To a child brought up with computers it is obvious that, no matter how good the simulation you see of your friend's face on the monitor, it is not actually your friend. It is clear that a movie starring a computer-generated character does not have the emotional impact of a film in which the star is a living person, someone to like or dislike in real life as much as on the screen. It will be manifestly clear to the generation of tomorrow that, however much data and information you amass, you do not necessarily acquire knowledge and wisdom, just as we know that, however many plastic beads a factory churns out, it is not making pearls. The difference between the ability of a machine to process data and a human to create ideas will be instinctively obvious to the younger generation.

Since her childhood in pre-industrial Krakow, Janina Suchorzewska, the protagonist of our Prologue, had moved with the times, as mankind harnessed electric power, made the car and the plane, split the atom, created radio and television, discovered DNA, developed antibiotics and heart surgery, took to space, and learned how it could blast itself out of existence. Yet the memory I have of her is of a woman in her nineties, decades into retirement, sitting at her piano and surrounded by her paintings, the fervour still there but now expressed through the keys. Her life had returned full circle, slowed again to the pace of the nineteenth century, to the music and art and ideas of her youth.

If living through ninety years of the most tumultuous technological and societal changes the world has known left her and her contemporaries with their values and ideals undiminished, we can have every hope that today's technical revolution will likewise leave our human qualities intact. Technology is changing the stage on which we are acting out the drama of human life, but it is not replacing the actors, who are still people – biologically, intellectually and emotionally the same people as those of ancient Greece, the European Renaissance or the American Revolution.

We have every hope that in the twenty-first century Emma will not have to face conflicts as appalling as those experienced by Janina in the twentieth. But in other respects we can expect that her life will be every bit as full of human drama, the logic of computers not subtracting a jot from the range of emotions and drives which will shape the behaviour of the people around her.

Further Reading

On the shelves around my desk are the numerous books I have pored over while condensing my thoughts into this briefest of summaries of the digital age. Here is a selection of them, with a phrase or two about each to help the reader share my delving.

My favourite among the experts' views of the state of the new technology, thorough and thoughtful but not heavy going, is

> *What Will Be: How the New World of Information Will Change Our Lives*, by Michael Dertouzos (Piatkus, 1997).

The author is head of MIT's computer-science laboratory. For a quicker read, by an unabashed enthusiast of the digital future there is

Being Digital, by Nicholas Negroponte (Hodder & Stoughton, 1995).

Hot on its heels, in somewhat similar vein, came

The Road Ahead, by Bill Gates (Viking, 1995),

which, though very readable, does not do justice to Gates's legendary intellect.

For a more extended tour of the fundamentals of the technology you could start with

Crystal Fire: The Birth of the Information Age, by Michael Riordan and Lillian Hoddeson (Norton, 1997),

which is a lively account of the invention of the transistor and the chip. For coverage of fuzzy logic and nano-technology I would recommend, respectively,

Fuzzy Logic, by Daniel McNeill and Paul Freiberger (Touchstone, Simon & Schuster, 1994)

and

Nano, by Ed Regis (Bantam Press, 1995).

The brief history of the World Wide Web is told through the eyes of the people who created it in

Architects of the Web: A Thousand Days that Built the Future of Business, by Robert H. Reid (John Wiley, 1997).

To develop a feel for the Internet as a societal phenomenon, the reader will find in

Release 2.0: A Design for Living in the Digital Age, by Esther Dyson (Broadway Books/Viking, 1997)

a guide for citizens, parents and policy-makers by a leading insider. A sharp but fair critique of the habit corporate executives have of spending blindly on computers, plus a good deal of advice on how to spend more carefully, is given in

The Squandered Computer: Evaluating the Business Alignment of Information Technologies, by Paul Strassmann (Information Economics Press, 1997).

A much more enthusiastic corporate business perspective can be found in

Blur: The Speed of Change in the Connected Economy, by Stan Davis and Christopher Meyer (Addison–Wesley/Capstone, 1998).

The view that computers must increasingly be modelled around the principles of biological organisms is set out in the interestingly written

Out of Control: The New Biology of Machines, by Kevin Kelly (Fourth Estate, 1994).

Another angle can be found in

After Thought: The Computer Challenge to Human Intelligence, by James Bailey (Basic Books, 1996).

The views of many specialists on how close we can get to the thinking computer are contained in this collection of essays:

HAL's Legacy: 2001's Computer as Dream and Reality, edited by David G. Stork (MIT Press, 1997).

The viewpoint of a scientist concerning the 'humans are fundamentally unique' school is summarized in

The Large, the Small and the Human Mind, by Roger Penrose (Cambridge University Press, 1997).

Anyone wishing to engage in discussion about living matter must nowadays be comfortable with the basics of DNA, as treated in a text such as

Signs of Life: The Language and Meanings of DNA, by Robert Pollack (Penguin Books, 1994).

The idea that this scientific discovery relegates human existence to no more than a statistical accident is put succinctly in

River Out of Eden, by Richard Dawkins (Phoenix, 1995);

love him or hate him, you have to agree that Dawkins puts his case well.

The difficult puzzle of how people – or computers? – can have consciousness is a big topic among philosophers of the digital age. A readable contribution is

Kinds of Minds: Towards an Understanding of Consciousness, by Daniel C. Dennett (Weidenfeld & Nicolson, 1996).

The subject is tackled by the co-discoverer of DNA himself in

The Astonishing Hypothesis: The Scientific Search for the Soul, by Francis Crick (Simon & Schuster, 1995).

For a broad and highly accessible coverage of the workings of the human brain I would recommend

How the Mind Works, by Steven Pinker (Allen Lane/Penguin Press, 1997).

Looking further ahead, a synthesis of the views of 150 scientists as to what future technology will bring is contained in

Visions: How Science Will Revolutionize the Twenty-first Century, by Michio Kaku (Oxford University Press, 1998).

That book will not disappoint those who like their futurist reading to bristle with technological marvels.

And, finally, for those intrigued by ISAAC in Chapter 1, the full story is in

Gridiron, by Philip Kerr (Chatto & Windus, Random House, 1995).

Index